P9-CDX-112

The Complete
Handbook of
Robotics

Dedication

TO

Cindie and Brandon

and

Tommy and Valerie

and

Eddie IV

who

love

Robots.

The Complete Handbook of Robotics

By Edward L. Safford, Jr.

TAB BOOKS Inc.

BLUE RIDGE SUMMIT, PA. 17214

FIRST EDITION

FIFTH PRINTING

Printed in the United States of America

Reproduction or publication of the content in any manner, without express permission of the publisher, is prohibited. No liability is assumed with respect to the use of the information herein.

Copyright © 1978 by TAB BOOKS Inc.

Library of Congress Cataloging in Publication Data

Safford, Edward L.
    The complete handbook of robotics.

    Includes index.
    1. Automata. I. Title.
TJ211.S23   629.8'92   78-10692
ISBN 0-8306-9872-8
ISBN 0-8306-1071-5 pbk.

101170

Cover photo is the "Grivet," manufactured by Gallaher Research, Inc.

BELMONT COLLEGE LIBRARY

TJ
211
,S23

Other TAB books by the author:

No. 74 *Model Radio Control*
No. 122 *Advanced Radio Control*
No. 135 *Radio Control Manual 2nd Edition*
No. 523 *Guide to Radio-TV Broadcast Engineering Practices*
No. 575 *Modern Radar Theory, Operation & Maintenance*
No. 631 *Aviation Electrioncs Handbook*
No. 671 *Electrical Wiring and Lighting for Home and Office*
No. 825 *Flying Model Airplane & Helicopters by Radio Control*
No. 939 *Handbook of Marine Electronic and Electrical Systems*
No. 959 *CBer's Manual of SSB*

PREFACE

How delightful to meet you again (or for the first time) through the pages of this book. We express our gratitude to you for your acceptance of our other books and hope that you will find this one even more exciting, interesting, and informative than those we have written previously. This one looks at the many facets of robots as they appear in all their various forms, so we have called it the Complete Handbook of Robotics.

The word robotics relates to a fascinating subject. Robotics is the fancy name for robot technology. Other names such as androids (droids for short), humanoids, and others are also used for these machines. Some have a humanlike form, some do not. But regardless of the name (and we have some others to explain later) these are machines able to do those things *we* don't want to or cannot do but which must be done somehow to make our life more pleasant and satisfying.

In this work we will try to find out why and how robots are put together and how they work. We want to know what they can do and what they cannot do, and especially what makes them able to do the things they *can* do. We do not intend to dwell on the construction of a single type robot, (for that, see TAB book 184), rather we want to spend our time on the fundamentals which apply to any and all kinds of machines falling in the robot category. Once we know these principles we then can experiment and use our own ingenuity to build and make operable whatever robot will most satisfactorily fit our pleasures and needs.

We ask you to remember that we already have many kinds of robots performing tasks for us. They do not appear in the fanciful humanoid form that so stirs the imagination, but they do exist. We will consider them in the forthcoming pages as well as that marvelous fellow, the domesticated android, as he has been termed. This fellow, we are told, is able to answer the front door, vacuum the carpets, stand guard day and night over the household, teach the children and play games with them, and perform other tasks for which it has been programmed.

This *is* the dawning age of robotics. The advances in science and technology have been so tremendous that now we have computers, the foundation of the robot's brain system, with powerful elements no larger than a postage stamp. The robot brain is considered in Chapter 6.

In any event it would seem that humanoid toys are now becoming more popular. Even these can perform various functions for your amazement and amusement. And with the general public's acceptance of the more complicated types of radio-electronic gear because of the CB explosion, it is inevitable that there will be a desire to own and have on hand one's own robot. Computer hobbyists will love the challenge it presents, as will all electromechanical hobby buffs. After all, what finer gift to the man or woman who has everything than a well mannered, tireless, hard working (for no pay) robot.

Your own fancy and imagination can flesh out details of what the future holds, perhaps more than our few words here. So, let us progress to a short discussion of what we have attempted to include in this work. First, we want to examine the general background of robotics and perhaps re-evaluate some types of robots. We will then progress to what robots can do and what they cannot do. We must be aware of the realistic capabilities and limitations of robots if we are going to use them intelligently.

A robot, like a human, must have *sensors*. What a magic word that is. It conjures up images of devices able to examine and evaluate a variety of phenomena. So we must consider sensors and learn something about them.

Also, a robot, like a human, must have some source of power to make it function. This must be considered in some detail, and the methods by which this flow of power may be made adequate and unceasing must be thought about.

Did you know that robots of all varieties, are made up of systems? There are internal systems, and external systems operating in conjunction with the internal systems. Not all robots have all systems. But the general case will consider a range of systems employed in elementary and advanced robots. So, we too should make this consideration.

Oh, boy! How about robot brains. What a huge and fascinating and unknown subject this can be. Yes, we must consider the many possibilities here. And we will learn some new terms and definitions for our vocabulary, as well as some larger blocks of fundamental knowledge. We will try to do this without the pain of complex explanation. From computer buffs who are very knowledgeable in this subject area, we ask your indulgence. We want to avoid great mathematical complexity. Incidentally, Kryton, the Chief-Emperor-King Robot of them all, tells me math is a must subject if you really want to become an electron brother to the extended-life-form robot, and android. So don't shy away from this subject. We believe you'll find it pleasant, fun, imaginative, and worthwhile.

Back to sensors again? You bet. It's a vital subject. How else is the robot to know it is *you* talking, or to see where people or objects are. Sensors, like our own seeing and hearing, touching, tasting and odor-sensing parts are very necessary to robots. We will go into more detail about them because of this. Is it possible that humans can sense electric and magnetic fields? Do we really have many other sensors in our bodies we really aren't aware of, but which give us that uneasy or marvelous feeling at times? Let's think about that. Then, of course, we want to consider some basics of robot design. There are many types of robots, and there is a whole range of capabilities. We all know about robots used in medical applications, in space technology, and in our homes. Let's have some exciting moments looking these over and then considering some of the basic design principles of various types of robots.

Klatu is a Robot, who, we understand, is the chief honcho of Quasar industries. His designer, Tony Reichlet, and I spent much time (and pleasant times they were too) talking about this. Now Klatu, whom I met, said—now he didn't come right out and say it in words, but he hinted—with a flash of his round illuminated head/dome, that we should include some *slight* reference to the manner in which the magicians of engineering design robot systems. That means some symbology and manipulation of it, and some "magic." We will have a chapter on the Black Art of Servo-Mechanisms. Just in case your membership in the Magic Mathematicians club expired, and you don't want to be advised of this dark science, you *could* omit this chapter without damage to your knowledge progression as a whole. But for some—who do hold high office in the world of magicians (and trolls and gnomes and muses) we offer something for their consideration in Chapter 8.

We must, of course, look at the actual construction of various robots, and then—unless we want to have robots running amok—must consider the test, adjustment, maintenance, and repair of these units. It is unfortunate that we don't have a Star Ship device which could effect repairs immediately through some invisible beam or force when the injured robot is placed on a pallet underneath the device. But we don't have these *yet*, so we do the best we can with what we have—ourselves.

As always, we acknowledge with humility and sincere appreciation the wonderful assistance received from those manufacturers whose names appear herein. And we thank those many, many Robot Masters with whom we have discussed this fascinating subject. Their enthusiasm and faith in what the future will bring in the way of androids (some say definitely) is marvelous to behold. Our gratitude and thanks to them.

Thank you kind reader for going through these pages with me, and thank you for considering this and our other books and giving them such fine acceptance. As we said in the first lines of this part of the work, it certainly is nice to meet you *again* in these pages if you have previously considered our other books and it's so nice to meet you here *for the first time* if you are a new reader of our efforts.

Finally we need to thank our publisher and our fine editors and the many staff members at TAB who put in such long and arduous hours making a good book out of a lengthy and sometimes verbose manuscript. There is nothing more heart warming than a wonderful relationship between author and publisher.

Now with Kryton and Klatu and all the gnomes and elves, and trolls and wizards and magicians standing by, being ever critical of everything we do, we begin the examination, study, and evaluation of this wonderful and delightful subject—robots.

Edward L. Safford, Jr.

CONTENTS

CHAPTER 1
ROBOTS, ROBOTS EVERYWHERE

Oh, I could tell you some stories, and I have in the last chapter of a previous TAB work called *Advanced Radio Control,* about the beginnings of robots as a functioning kind of machine. Fascinating! Yes, all of that and more.

WHAT IS A ROBOT

Now, here we go on a more-detailed examination and explanation of robots, which, to coin a definition, are fully automated machines which may respond to external stimuli as well as to internal commands which have been prerecorded. It is important to note that we have here the term "robot," which is different from android, or droid for short, or from humanoid, another term associated with these machines.

We are using the word machine, and this may be offensive to some, especially Kryton who thinks he is almost human—in fact, sometimes I do believe he thinks the normal situation in the universe is reversed and he is the human and we are the machines. His logic tells him this, so he informs me!

SOME TERMS AND DEFINITIONS

It probably is important that we come up with some terms and definitions as we discuss this large world of electromechanical automated devices. Robot, as a single term, may not be definitive enough for our detailed consideration of what takes place and how it happens.

Let's start with the dictionary. The word robot is defined as coming from the Czech word robotnik, which means serf (or slave), or robata, meaning compulsory service. Also, the word robotisi means to drudge, which means, essentially, an unthinking kind of *work*. The definition also says that it means any of the man-like mechanical beings in Karl Capek's play R.U.R. (Rossums Universal Robots) who were built to do manual work for human beings. Incidentally, the robots improved so much that they took over the factory and presumably the world! Quite a play, and probably the first mention of the word robot to mean a mechanical automated machine. Finally, a robot is defined as a *person* who acts mechanically without ever thinking for himself. Know anyone meeting this definition? There is no argument contrary to the fact that all robots, or whatever we may call the progressions of this device, belong to the world of automation. We define that as meaning the automatic operation of, or accomplishment of some events, or physical motions, or measurements, or issuance of some phenomena without the benefit of *direct* human control. Of course it was the *human* brain which put the device together and determined just when and how it would do all these things.

Because the art, and we will insist it is a kind of art as well as a science, of robotry has come a long way since Capek's play R.U.R., and will continue to advance in the years to come, more definitions will be needed to specify the various types and levels and capabilities of operation of the present and future robots. Although it is certain that the word robot will always be with us and have its own general and distinct connotation.

I am thinking right now of a movie—science fiction type— where two robots were guided and operated by a giant underground computer. They finally had to be destroyed, at great risk to the hero of course, because they ran amok. They were named Gog and Magog. It is interesting that these names stand for evil personifications in the biblical Book of Revelations. But the general fear, which may be generated, that robots will run amok just isn't founded on fact, at least at the present writing.

But, back to the development here, we need to consider the current labels attached to various types of automated machines falling into the category of our discussion. Kryton, and no one dare doubt his word, tells me that some definitions are as follows:

machine-automated—May be any machine that can do something without human control. A very general term used in the world of robotics. Can mean the most basic and elementary kind of robot.

robot—A general term which stands for an automated unit which does have some kind of human symbolization in its operation or physical structure. It may not look human, but it seems to be able to perform *human type functions* and actions. A computer may perform something similar to the human brain function. A car-wash machine performs a similar-to-human mechanical function on a larger than human capability scale.

humanoid—A machine which has more of the human characteristics in its physical appearance. Although the appearance may be symbolical there is an allusion to arms and legs and body and head and arms. May be large or small depending upon its task performing requirement.

android—A human appearing type of machine which duplicates in appearance and some actions the human body and its functions. May converse, and does respond to external stimuli with pre-programmed responses. This is the type visioned by sci fi writers as the companion to the hero in many far-out escapades.

cyborg—A biological type of robot which has all capabilities of the previous types of machines and also has some artificial intelligence, or thinking capability. The word is related to cybernetics, which is the study of the human brain and human nervous system, or a study of complex types of calculating or computing machines.

cybot—The machine counterpart of the biological cyborg.

cybert—This one we coin here to mean the most advanced type of robot possible to imagine, and with human appearance. It has all the basic mechanical talents; it responds to external stimuli, and it *learns by its own experiences*. It has a fantastic computer memory as part of its being.

Of course there are terms and definitions *you* might want to add as they come into our vocabulary as we use them at the appropriate time. We don't say that our little list of definitions is all inclusive nor the most accurate, but as Kryton says in his melodious voice monotone, "They'll do for the time being!"

WHY ROBOTS

It might be said that humans need automatic machines to do things which humans cannot do because of fatigue, strength limitations, speed limitations, precision limitations, environmental limitations or length of time of operation limitations. Of course we all know that humans have designed equipment to extend their own capabilities. For example, scuba diving gear extends human breathing for work or play in the non life-sustaining environment of the sea or other underwater depths. That basic idea may be used to design suits for use in toxic atmospheres where some product or process is contained. Space suits could probably be put in this category. A little thought will bring to mind countless other man-assisting devices currently in use, available, or on the drawing boards.

Man has made for himself a series of machines to extend his slight physical strength. Bulldozers, excavating equipment, mining equipment, and hydraulic presses all exert forces which man simply cannot produce through his own fleshly efforts. Some of his machines are so precise in measurement and control that they can detect variations of tens of thousands of an inch (or centimeter) in drilling and placement functions. Radar as well as television, night-vision scopes, and other devices extend man's vision beyond his own physical limitations.

Man is able to "touch" and "feel" beyond his own reach through the use of sensors which sample, translate, and relay information to him. Remember the Mars experiments? And, of course, modern computer technology has produced electronic machines able to perform "brain exercises" faster, more accurate, and more complex than man ever dared to think of being able to do by himself. Some have wondered if man will be placed, ultimately, in a reduced state of mental and physical capabilities just because he has, and will continue to invent so many electromechanical marvels to do everything for him.

Considering all this, then, there is a vital need for these machines, thus few ever argued this aspect of the technological development. Making the operations automatic also is vital in many situations. Think of just a few which make your life easier. The air-conditioning robot, the automatic dishwasher robot, the clothes washer-dryer robot, the automatic car wash robots, manufacturing process robots, and sorting and filing robots. You will likely think of more. Whenever there is a need to perform some function continuously, accurately, whether in a hostile environment or not, automated robots are considered a prime solution to the need. There does not seem to be a question as to their capabilities, which are being extended all the time as we humans learn to adapt the interface devices necessary between the controller and the controlled units.

ARE THEY TRUSTWORTHY

Probably the most basic question concerning the use of robots is the age old bugaboo of a malfunction. Stories have been written, horror stories I might add, telling of a robot "gone wild." Man always has this basic fear deep down in the folds of his own brain. Malfunctioning can range through a whole myriad of situations. The most basic is that the machine stops working. The most terrifying is that the machine works, but in some unaccountable manner to produce destructive results rather than pleasurable and constructive. We think now (advised by Kryton and Klatu, who say, incidentally, that all robots and members of that social order are really benevolent fellows) that no such problems will develop.

But we know of the self-destruct features man builds into robot machines in an attempt to do away with the basic malfunction fear. There are a range of operations which can be initiated when a robot doesn't do what it should do. These might be the blowing of a fuse, rendering the robot (or higher social-order machine) inoperative and harmless, to an actual self-destruct, such as must be activated on guided missiles and rockets if they stray from their prescribed and commanded path.

WHAT CAN THEY DO

How many robots do we now have? A whole world of *industrial robots* exists which go about their business of making things we want, quickly, efficiently, accurately, and economically. There are toy robots which, as time goes on, will perform more and more tasks in the fun world for us. There are security robots to monitor and maintain your home secure against interlopers and to warn against fire and malfunctions of other equipment within your home. People will come to think increasingly about robots, and our children will find more and more of them around in their future. Examine Fig. 1-1. What a combination of imagination and wit from some unknown electrician. With curiosity and perhaps anticipation aroused Tom Safford, Jr. studies this mechanical man.

We said we had occasion to discuss domestic robots with one of the leading authorities on this subject, Tony Reichelt of Quasar Industries. That, incidentally, is his manufacturing concern, which will be producing domestic robots, security robots, and sentry (security) robots in the future, he said. It was a most interesting discussion because Tony, admittedly biased on the subject, presented a glowing picture of future capabilities of these robot types, as he envisions them. Of particular interest was the paramedic robot.

What's a paramedic robot? That is a robot used in the medical profession to assist, as commanded, in giving treatment to certain patients, possibly those boys and girls who have been maltreated by adults in early childhood. They may consider the robot a toy, in a sense, and yet be someone whom they can speak to and confide in as a *non person*. Much might be learned through the use of this confidant.

There are other uses too. Mental problems which require constant monitoring, lonely people who need a "conversational companion" or the security a companion robot can provide, special problem cases of adults who require constant monitoring and/or help are other areas in which this kind of paramedic robot might be of great value. It was indeed marvelous to listen to the enthusiasm and vivacity with which Tony described the possibilities as he sees them. And who is to say these possibilities are not realizable?

16

Fig. 1-1. An electrician's imagination stimulates a future scientist.

So, Kryton reminds me, people know about robots. They have seen robots in movies, usually the mean and ugly kind, although there are a few who are "nice" and "good." The movie Star Wars brought into being the R2-D2 computerized robot who could speak a language, and his friendly android companion. Then, of course, there was Darth Vader.

Yes, everyone knows about robots of some type and can identify some other types if a little thinking is done on the subject. Remember, a robot is an electromechanical machine. It performs functions without human control once it is activated by human control. Or it may respond in accord with various stimuli received on its own sensors, once it has been activated by human control. Notice that we humans always want to have the power of "life and death" over the robot in any of its generic forms. That is, the power to turn it on and turn it off.

CREATING A ROBOT

Now, what's the big problem with making robots? We seem to have the scientific know-how and technical capability to do almost anything. What prevents us from having a lot of robots, humanoids, and androids around? Would you believe it if we state that it is the lack of a self-contained and accurate reference system which gives us one of the biggest headaches? It is. At least that is what Kryton insists, and as the chief-emperor-king cybert of them all, he ought to know. Also there are the "need" factors. Why do we need them? Do we *really* want them?

Reference Systems

We humans have marvelous reference systems. We have eyes to tell us where we are by visual observation. And we have a memory and comparison unit in our brains to inform us that what we see belongs to a certain physical location. How many times have you asked the question "Where are we?" It is then that we begin to *look around, trying to match up what we see with what we know* or with where we have been. Or we go to maps, which pictorially tell us where we have been and, supposedly, where we are going. Using the map usually requires some computation. We have to know in what direction we have been traveling and for how long. Or we need to know what winding or zig-zag direction and for how long we have traveled, to determine our displacement from our starting point.

If we have an aircraft, and sometimes in an automobile, and generally always on a boat, we find that map travel is related to that all important sensor the compass. It always knows where north is and tells us so. Even then we may wind around and swing back and forth trying to follow a set course related to the compass indication. Winds affect both aircraft and boats. Waves and currents also affect boats. Metal in and on the earth affects the directions given by the compass in a land vehicle, and may do so in an airplane. So there is "hunting" and swinging and other variations on the path we may follow using this kind of directional reference.

We bring this discussion up for consideration simply to indicate that obtaining a directional reference *is not easy*. And to have such a reference in a robot, humanoid, or android is one of the largest problems yet to be solved. Let's just consider going for a walk with a robot. If he is adjusted so that he will maintain a given distance to your side, then he will turn and move in accord with the *preset* command. He can accompany you everywhere, keeping the precise distance from your side, unless there is an obstacle in his path. Suppose you go through a door or gate in a fence and he is keeping his distance beside you. Will he fall back and follow behind you at the required distance, or will he go ahead on into the wall like the poor robot in Fig. 1-2. Ah, you say, let the robot have sensors which can override the basic direction preset into its computer type brain. If there is an obstacle, let it stop, make a noise, or move sideways until it no longer has the obstacle in front of it. That would be nice. But think for a moment of what difficulties this might present, based on the range at which it might have to detect such objects. If it detects them too close, it might move sideways quickly and run into *you*. If it detects them at long range, it might be moving sideways when it shouldn't. That would cause it to lag far behind you and then to follow a very erratic course trying to catch up. Think about it. You'll enjoy going over in your mind the limitations of going for a walk with your robot.

Consider having a robot in the home. There are tasks which it could do—and mother and wife will love it for accomplishing them. It never tires no matter how much drudgery is associated with the tasks. It is efficient, presumably quiet, and thorough. But how does it make its way around the house to do the necessary *preprogrammed* tasks. Let's take sweeping the floors. How does it solve the furniture problem. Does it avoid these items or does it move them and

Fig. 1-2. Walking with a robot.

return them to their original positions. Does it jerk or yank, the items or does it move them slowly and carefully? *How does it know where it is if it doesn't know where it has been?* Some wizards think that light and color contrast of walls and furnishings could establish a matching situation of the objects so the robot can tell where *it* is.

Let's consider a possible scientific development which would seem to verify this possibility.

There has been a General Electric development in a TV pickup device which causes an electron beam to scan across a computer card matrix so that it breaks down the viewing area into thousands—perhaps tens of thousands—of tiny squares, each of which may be connected to a further matrix in a computer memory bank. Thus when the "eye" scans an area, everything in that area, *from that position of the scanner*, can be identified by its light reflectivity and color. So the computer defines a shape for the objects. If, within the robot memory, a similar shape had been electronically preprogrammed, matching of the two shapes would establish an identification, and thus possibly the objects position with respect to everything else. Ah! You guessed one difficulty. The shape or appearance of anything changes in accord with the viewing angle; thus, the memory would have to store a fantastic number of possibilities for the robot to know precisely where it is, from a given picture of things ahead or around it. Think about it.

But the possibility is real. And, if one were to assume that the robot always came around in a given direction, he (or she) might be able to have a fixed visual identification stored in memory banks which would always be the same. Thus the problem might be solved in this manner.

Personalities

We have mentioned the robot in connection with the pronouns he and she. This brings up a very important aspect of home robots. It seems that it is very important to match the personality of the domestic robot to the personalities of the members of the household in which the robot would serve. To do this a personality profile study of the family may be necessary. Based on this would be the robot's responses to commands or interrogations by family members. It is also possible that in some cases, say with young children whom the robot is watching or being a "nanny" for, "he" should possibly be a "she" in appearance, voice, and mannerisms. So the robot is provided with feminine voice characteristics and perhaps mannerisms which are compatible with the personality she must project. Children are used to mamma telling, instructing, explaining, and giving them dicipline, perhaps more so than the fathers. So for the young children to whom the robot will be a guardian, teacher, and companion, the identity with "mamma" makes it more acceptable and authoritarian than would be possible otherwise.

Personality characteristics may also be a vital key when using robots to attend the sick, aged, and mentally maladjusted persons. The proper personality projected by the robot might be able to establish communication or adherence to requirements both physical and mental when the presence of an actual person is impossible or is not considered to be safe or feasible.

As of this writing, robots have personality. The Quasar SPA (sales promotional android) in Fig. 1-3 has become a debonaire man about town and is apparently attractive to women, if the models attention can be thus interpreted. That he has a personality, as evidenced by his costume, there can be no question. With voice and voice mannerisms to match he can become so real that you wonder if perhaps there isn't a little man hiding inside his cone-shaped body.

THE ROBOT MASTER

This robot has a weight of about 180 pounds, stands 5-foot tall, and has a round, plastic head with wires or light conductors inside which flash and glow as he speaks, giving a vibrant and real-life mystique. He has a cone-shaped body because this is the strongest and probably the best balanced structure possible. The Robot Master assured us that he can carry a 45 pound boy on one arm without the least danger of toppling. The cone is continuous and of stainless steel so that it can withstand an unbelievable amount of physical abuse. This is perhaps a vital note in considering the use of a robot around the home where children may experimentally pound on it.

The arms are covered with a flexible accordian type covering to permit movement in normal arm directions. When the arms move the motion is smooth and fast, but not abrupt. The arms seems to "glide" to their final preprogrammed position naturally. Inside the robot is the maze of transistors and microcircuit boards, about 8 by 10 inches in size. These constitute its control circuits, memory, and communication circuits. Electric motors furnish the motive power for the three wheels, making it a triped type machine.

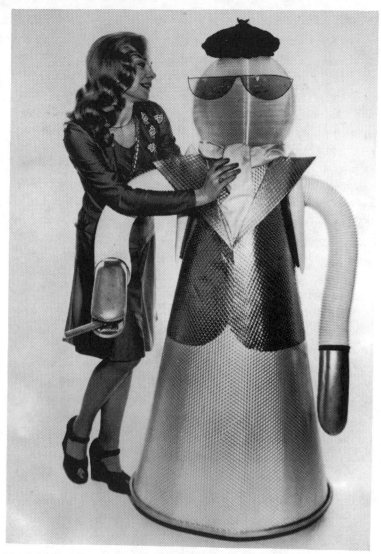

Fig. 1-3. The robot projects a personality. Courtesy Quasar.

While the claims by the Robot Master for this kind of robot (Fig. 1-4) and its future are vast, there are many other authorities engaged in the various fields of automation who believe these claims may be exaggerated and that there is some kind of trickery involved in making Klatu and the SPAs do what they are doing as of this writing.

THE POSSIBILITIES

In this book we will be examining some of the scientific and engineering possibilities and thus may come to some conclusions ourselves as to what can and cannot be done in our marvelous age of exploding science and technology. We do know that there are computers capable of almost unbelievable operations which can be programmed rather simply from desk-type keyboards. We know of a large number of studies devoted to voice recognition and some success has been obtained in developing a voice writing typewriter, where you just speak to the machine and it prints out the input. We know it is not inconceivable that Klatu can be programmed to do many operations simply by using a small hand-held (like a minicomputer) programmer. This, incidentally, is used by the people at Quasar who operate these robots for fun and profit.

There are possibly other developments which are highly guarded secrets (even to the CIA and KGB) about robots, which we will find out about when they are put on the market. We asked Quasar Industries for some technical information on their SPAs, but they informed us that although patents are pending on the series they still guard the basic information jealously, for with but a small change in concepts other manufacturers may use the ideas under new patents. Such is the case for all manufacturers, scientists, and engineers. They live in a world which is a constant nightmare because their "ultimate" breakthrough may be stolen and exploited. They lose money and prestige, just to mention a part of the penalty of loss.

Off the Shelf Hardware

If you are a technical detective, trained in the search for and evaluation of new electronic and mechanical developments, there is a good chance that you will begin to think of off-the-shelf items able to be put together in a machine which then could be called a robot or possibly even an android. Let's take, for example, the simple telephone answering machine currently on the market in a variety of sizes and configurations.

A telephone answering device selling for about $250 has within it several features of importance to a developer of robots. One is that it *is* a program type unit. It *will* record any information which comes into it via a microphone or wire attachment. We extend this idea to say that with a little adaptation (of another unit currently available, a small radio communications set such as your CB) this unit can also record messages or, more importantly, instructions via the air waves.

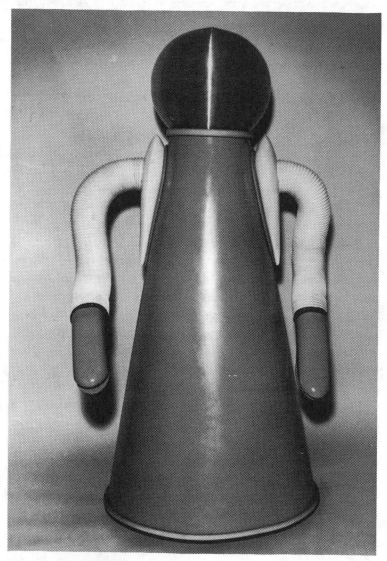

Fig. 1-4. Is this the future domestic robot?

Second, the unit remains off until it is activated by a voice input or the equivalent of the telephone ringing impulse or signal. When it does run, it may first give you (the caller) some information such as "Please state your name and the purpose of your call." Then it shifts its controls to record your information. New? Not really. Telephone answering devices have been around for a long tme in a variety of models, some large, and some small, and for our purposes here, the smaller the better.

Your Own Butler

Now consider some uses for this device in a robot which has as one of its tasks the answering of the front door. Let's leave the method, or *a* method, by which it might be alerted to the fact that someone is at the door aside for a moment. We'll also skip over how the door might be opened by it. Consider just now that the robot has sensed a person is at the door, that it is open, and that it is face to face with the visitor. Now, through a sensor or a timed circuit response, the voice recording circuit in the robot is activated. The first thing it says is "State your name, business, and whom you want to see in this household." Then it becomes silent waiting for the message to follow. What does the visitor do? Well, if they haven't fainted from fright, after the first gasp of astonishment, they respond by giving the message and probably feel very self-conscious standing there talking to a dummy machine which is confronting them. Perhaps they would be more intrigued by it if the robot has some-what of a humanoid appearance. In any event, they respond with some sound even if it is only a scream of fright. That gets properly recorded in the machine, incidentally, and can be used in future programming to make the robot more acceptable to people. If the first hundred who confront it are merely frightened, perhaps a new appearance or a softer approach saying something like "Do not scream. Do not be afraid. I am a robot, but I am harmless (at the moment). Please state your name and business at this address. Thank you." That might be an approach which would be more conducive to a nice reply. Of course you might also have to have the robot say "Please do not be self-conscious about talking to me," or something like that.

So what happens next? Well, the pause when your voice or the visitor's voice stops coming into the machine can be a command to some other robot machinery to make it talk again. Here, perhaps, two recoding units might be used. One activating sound from the robot, the other recording what the visitor has to say. But in any event, the silent moment causes the other recording machine to state a prerecorded message like the first, but with different information. It might be "Thank you. Please enter and follow me." Whereupon the robot turns and, following another prerecorded instruction makes a series of positive, definite movements involving mobility as it leads the visitor into the room. Then, with a series of arm movements to make it seem more humanlike, points to a chair or couch where the visitor can sit down. It then says "Please sit down. My master will come presently." Then, while the visitor sits in amazed silence it moves away, again under a prerecorded instruction, to where it can deliver the door message to you and then return to its station at the doorway.

Of course, the robot may be programmed to entertain the guest until you appear, the message as to whom and why they are there having been transmitted to you via radio when the visitor first spoke at the doorway. The scenes could be as shown in Fig. 1-5 (a, b, c, d).

Impossible? Not at all. Combining mobility features with the proper use of off-the-shelf voice recorders in a robot makes it possible. The recorders, we must admit, might have to be modified somewhat. And preset, programmed mobility circuit(s) developed, which we will discuss more in a later chapter. Fascinating idea isn't it.

Klatu, of Quasar Industries, is said to answer the door and lead guests into the house and speak to them. So it is possible, but take note that the manner in which we described the possibility *may not at all* be related to the manner in which this is said to be accomplished by that robot. We are just discussing the possible use of some off-the-shelf items to make robots.

FEAR OF ROBOTS

That brings up again the interesting consideration as to why there aren't more domestic robots and why they aren't used more in our current world. As of this writing people would probably rebel at the cost—estimated at $4000 and up. And many people are just a

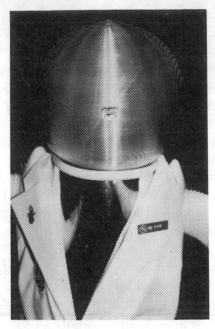

Fig. 1-5(a). Hello there, I'm Klatu. GE calls me Sam.

Fig. 1-5(b). I'll just join this group.

little afraid of these machines anyway, one that does something without a direct human hand on the controls. Fear, possibly, is one reason why people haven't demanded more of these home type robots than they now have. Yet good and benign robots have become quite important. One movie, Star Wars, had both. Threepio and R-2 D-2 were the *emotionless* (now catch that word and don't let it out of your memory banks because we later want to come back to what it means in Robotics) machine heros of the movie. Darth Vader, probably the one you'll most remember, personified evil and is the kind that may make us all, as humans, fear the technological development which would make humanoids or androids or the cybert robots, something to stay away from.

But the robots cannot think. They can *learn* from their experiences and thus do things better because they learn from their mistakes. But they cannot have imaginations. Science fiction writers love to bring out the possibility though. They speculate the machines will become so complex, so refined, and so filled with thousands of postage-stamp-size electronic miracles that somehow, some way, sometime, they might begin to *really think*. And of course the writer wants to give his readers a thrill, so he always has the device begin to have diabolical thoughts and become an evil machine. Well it *might* happen sometime in the *far distant future*, when we really know what thinking is. But chances are that it won't be our kind of thinking or imagination that can be programmed into a computer brain—ever. We know we may get a lot of disagreement on that statement, but we'll stick by it.

When people learn to accept robots for what they are, and people means you and me, we can begin to use them to more of an advantage. It will be a process of gradual acceptance, but it will come just as certainly as the automobile, the airplane, the satellite and the space ship did. They will arrive just as certainly as radio, television, and pocket-size computers did. And they will supply just as much fun and satisfaction as those other devices. The technology of robots is just beginning, in the sense that they are now called *robots* and not automated machines or mechanical devices or self-contained controllers, and so forth. They will move on their own. They will perform their tasks. And we will love them for it. They will have gracious

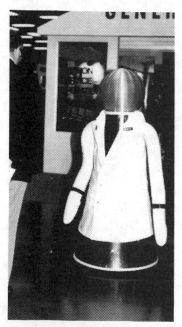

Fig. 1-5(c). Wanna talk or rassle?

Fig. 1-5(d). As I see it...

personalities adapted to us as individuals, and we will grow fond of them and love to have them around. See Fig. 1-6.

MOBILITY

Consider mobility for a moment. In this general discussion we need to come back to basics and think what mobility means in the robot world. First of all there are some terms, a biped, a triped, and I suppose, a quadraped, which simply mean a two- three-, and four-wheeled device, in that order. Klatu is a bubble-headed, cone-bodied, accordian-armed triped. Look back at Fig. 1-4. You don't see the wheels underneath his metal skirt, but they are there, rubber-tired wheels which give him mobility and steering capability.

With this in mind for an actual working robot, let us first answer the question, "Why not have it walk as we do?" And a good question that is. With mobility derived from wheels, the robot cannot climb stairs, may not be able to move over rough ground, and will likely hang up on small obstacles in its path. Yes, there might be some advantages to the walking idea if it turns out to be practical and feasible.

WALKING

What, then, are the objections to the walking concept? Let's consider first how we walk. Did you know that we unbalance ourselves in the direction we want to go and then quickly move our foot out to rebalance our bodies? 'Tis true. If we just walk slowly, the unbalance is slight. If we run or move quickly, then the unbalance is great. Wanna prove the point? Then have someone hold your feet still while you think of walking (and try to do so). You have to readjust your muscles quickly to keep from falling on your face. So that, right away, is one objection for walking locomotion for a robot, at least an objection to the two-legged walking concept. We have enough trouble trying to keep the robot upright with all the other things we want it to do without having to include additional compensation for an out-of-balance situation.

Now this doesn't mean that there couldn't be some type of system wherein the robot *could* have legs and walk and still have its balance. We are reminded of an article some time back telling of the military's interest in developing a vehicle able to maneuver over very

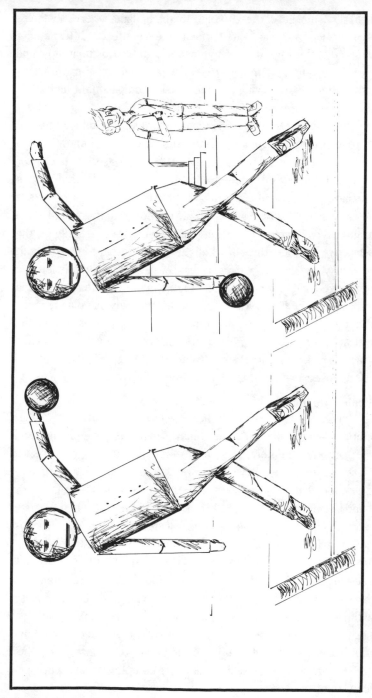

Fig. 1-6. Robots can be adjusted.

rough terrain where there were no roads, and where obstacles almost as large as small homes might be encountered. It turned out that some of the wizards we have in our country (fortunately) came up with a quadraped mount resembling a giant spider with four outstretched legs carrying a small boxlike compartment high up in the center of the thing, Fig. 1-7.

It was shown that this *could* maneuver over the difficult terrain that posed the initial problem. If it was used, we do not know. We do know that because of its slow speed it was a "sitting duck" for well-aimed artillery or rockets, and the motion could possibly have immobilized the operator from "sea sickness."

Discussing this machine is simply to show that there might be a walking robot developed, although it wouldn't neccessarily be the machine we so fondly think of when the name android comes to mind.

But, you ask, "How did the thing walk?" It raised one leg at a time, moved and extended it, and then when it reached the ground ahead, raised the body. Then another leg was advanced and moved forward till it could repeat the lifting and moving process. Only one leg at a time was off the ground. So three legs always maintained balance. Think about it. Play with the idea. You can picture how it was supposed to—and for all we know—did work.

Wheeled Mobility

But back to wheeled mobility. This is a swift, sure, quiet method of locomotion. It doesn't require an unbalance situation. And since a robot will likely perform its tasks in an area where such mobility could be used, wheels offer the best choice at present. Remember that large wheels can run over small roughness or ground irregularities, and smaller wheels can be used when the path is very smooth. This seems to be a sure first step to solve this fundamental problem of self-mobility for our automated, self-reliant machine, the robot. If just one wheel is turned for steering, say on a three-wheeled robot, this task can be simplified. The resistance to unbalance can be good, with proper location of the center of mass and the spread of the three support points. We are thinking of triped mobility now. Of course, four support points (four wheels) offer still better balance and stability, but the additional steering requirements might not be worth the additional cost, complexity, and perhaps, even size of the base.

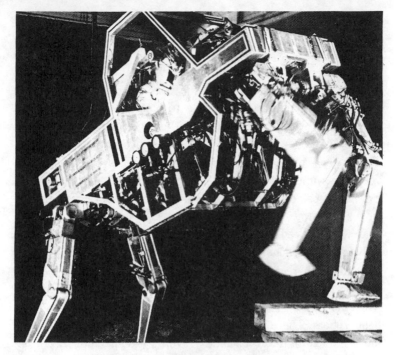

Fig. 1-7. A walking quadraped. Courtesy General Electric.

What are some of the problems concerned with the triped? Well, first you must realize that we want a smooth movement, from stop and very slow to as fast as the unit can be programmed or commanded to move. Some units developed by Quasar are said to move at about 20 mph, some are planned for up to 30 mph. That is pretty fast. But think again of the smoothness requirement. Of course we don't want jerking starts and stops. But we want full movement power available at almost stall speeds, a major consideration. Most electric motors which might be used for robot movements are controlled by variable resistances in their primary lines. And they have reduced power when slowed down. So just any kind of speed control probably won't be the proper solution for what we want. Again, we want full power, or torque, at any speed, and absolute smoothness from zero to full speed. Wow! We will have to examine this in more detail in a later chapter.

The movements of arms, body, and of course, the head could also require the same kind of control, except that these will probably have a fixed speed. Thus the motor can be adjusted for the required torque and smoothness at that speed. Starting and stopping movement so that it is accomplished without jerking is an important consideration with any part of the machine. It *could* turn out that hydraulics or pneumatics might be used simply because of this requirement. It is always possible to have smooth and powerful movements with these drives without jerking. You see this whenever you have your car lifted by a service station grease rack. The power is usually an air controlled hydraulic system. Of course you have witnessed the slight bump as the rack reaches its final extended position, and the plunger in the cylinder physically impinges at the top of the cylinder. But that can be compensated for in a machine, if desired. In a robot it is considered important to eliminate all bumps in any kind of motion.

Today there are more fixed robots doing important jobs in industry than we really imagine. We accept these automatic devices when they are presented to us as such and don't think about them being robots. There is a world-wide effort to expand the development and use of these machines, even if it sometimes means antagonism from those who fear loss of jobs or their lives being taken over by a mass of machinery and electronics. Computers have been developed to such an extent that almost everything in our lives involves a relationship to these machines. A computer may not move in a physical sense as do the robots we think about. But they have much movement in an *electronics sense* as they process and store information, causing motion in other machines which record, print, or process the output data.

THE ROBOT "BRAIN"

Aside from the physical construction and drive equipment, the most complex assembly in a robot is the computer. It must have some means of storing information and of acting on commands which can be repeated over and over when it is activated. It must have a selective system directing the motion systems to perform preprogrammed (or unprogrammed) actions. The selective system must know which commands have not been accomplished.

mputers have come a long way since the '60s. Now you can
rs built into a lead pencil. You can find little computers

having faces resembling owls, clowns, or other familiar characters. They invite youngsters to try their skill at mathematic solutions or computerized games. These little Nova type devices will tell a child when his answer is wrong or right, thus providing learning instructions.

No one is afraid of these computers, or "brains." Perhaps we should not be afraid of larger computers either, which may teach more of that subject and, indeed, many more subjects pertinent to the child's education. It might turn out that instruction by a nonperson having *infinite patience*, never becoming aggravated, always having the correct answers, never growing tired of repeating, might be one method of quicker learning. Remember that robot teachers *might* well be made indestructible and carry a large wallop for disciplinary purposes.

"Of course," you say, "it is true that we now have some computerized learning centers, and children do use them and they are effective. But they are costly and large and do not give the children the same group interaction they get from classroom studies and recitations."

The robot teacher may solve this problem. It can work with one or several of the students at a time and may be able to take questions from each, providing answers in a logical, sequential manner. The robot, as a teacher, seems to be a vital and lucrative area of investigation. Of course this means it must have a complex computer allowing it to do more than just move around performing ordinary tasks.

Thus we come to the idea that there will be "wheels within wheels," as the old saying goes. There will be systems within systems and functions within functions in order to obtain *full* realization of this kind of automaton's capabilities. Now there's a word. Automaton. It means automatic machine. During research, we found literature usually listed under either automaton or manipulator. Then, under these general headings, we found the word robot. We did not find android or humanoid or cyborg or cybert listed. We have to admit these terms belong more to the field of science fiction, movies and TV at present. The precise (and sometimes dull—but brilliant) engineers—who would insist that a piece of wood be called a piece of pine, cedar, or oak—favor the names automaton and manipulator. They think these are more realistic descriptions of what the machines purpose is or how it functions, perhaps.

ELEKTRO AND SPARKO

But let's dream a little for a moment and examine Fig. 1-8. Here we see the Westinghouse robots Elektro and Sparko. These are, perhaps how we imagine our future robots will look.

Let's examine the following information about them. They have been "brought back to life" after a retirement of 8 years. Elektro is now 11 years old although he is not as smart as an 11 year old boy. He has a vocabulary of 77 words, never gets tired or hungry, and speaks *only* when spoken to.

Elektro is large for his age, he stands 7 feet tall in his aluminum feet and has an 82-inch chest expansion. His chest always is expanded because, like the rest of his body, it is made of aluminum over a steel frame. His feet are 18 inches long and 9 inches wide. His energy comes from the nearest light socket.

This mightly automaton will never need a psychiatrist because his brain consists of 82 electrical relays. These relays operate his nervous system of motors, levers, gears, and chains to make him walk, talk, count numbers, smoke, distinguish colors, and salute. His spinal column and other nervous system contain hundreds of miles of wire.

Elektro can perform 26 motions in all, but he is a dullard in comparison to any man. There are 292 muscles in the human body, capable of producing thousands of different movements beyond the 500 most rudimentary motions. On the basis of Elektro's 260 pounds and 26 motions, an automaton would have to weight 5000 pounds to perform the most elementary human movements. But based on new technology and miniaturization, that weight is very much more than we would require now or will see in the future.

Elektro's 60 pound brain includes an electric eye, 82 electric relays, and signal lights. In order to direct the full 500 elementary motions, a new robot constructed the same way as Elektro would have to have 1026 relays, would weigh 1000 pounds, and occupy 108 cubic feet of space. But remember Elektro was designed many years ago. A new robot would take advantage of the latest technology.

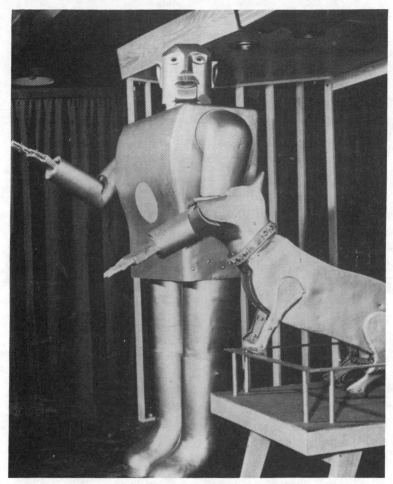

Fig. 1-8. Elektro and Sparko. Courtesy Westinghouse.

No matter how capable a mechanical man may be, it still must be "bossed" by a human. Elektro responds to commands spoken into a microphone. Each word sets up vibrations which are converted into electrical impulses. These impulses operate relays in Elektro's brain, controlling his 11 motors.

Talking to Elektro is like dialing an automatic telephone, using light impulses instead of numbers to cause the relays to act. A series of words properly spaced select the movement Elektro is to make. Two-word commands start the action. One-word commands stop it. Four-word commands return all relays to their normal positions. It makes no difference what the words are as long as the proper number of words (electrical impulses) are produced. Two electric eyes—one with a green filter and one with a red filter—enable Elektro to identify red or green lights when they are flashed before his eyes. His walking is accomplished by a motor which drives the four rubber rollers under each foot.

Nine motors are required to operate the fingers, arms, head, and turntables for his talking. Another small motor works the bellows for Elektro's smoking. (He can stop easily by simply not energizing the motor. Wouldn't it be nice if we could do it so simply.) Elektro's talking is produced by recordings. He uses a lot of energy. If all the energy of his motors were applied to one task, one horsepower would be expended (550 foot pounds per second). Sparko, the mechanical dog, is operated by two motors, which are also controlled by voice commands.

COMPUTERS IN ROBOTS

When thinking of the computing aspect of the robot, it is interesting to think in terms of the microprocessors, memories, and other devices now on the shelves for the assembly of most any kind of computer. Many companies are producing computer kits intended for the general education and use of the public, people like you and me. So there is a technical revolution in process. The *computer*, in a larger sense than the hand calculator, will become a household word and "appliance." It will be as important as the hi fi, TV, and other such devices are today.

We have been informed that home computers will be used to plan meals, handle bill payments, balance the checkbook, maintain appointment schedules, and monitor and control heating and air conditioning. The latter would be nice, for sometimes we want to cool only one room. The computer would direct the cooling there and not expend the energy trying to cool every other room when just one room might need cooling (or heating).

Using a computer for home tasks will be quite common in the future. So it is not such a far step then to the idea of using a computer in a robot for the programming of its various tasks. In fact the   might not be in the robot per se; it might just control the   anywhere. Watch for this kind of advancement.

We had occasion to discuss the use of computers in robots with one of the men who designs the microprocessor and computer systems. In answer to our question as to whether he thought it would be feasible to have all the computer information for complete automation of a home robot with a small enough size, with enough reliability, and with ruggedness and simplicity of operation to be practical, he answered quickly and with enthusiasm, "Yes! There is no question about that. Items now available can easily meet this challenge."

We then digressed to the amount of computer information necessary for the robot to steer itself around the house and perhaps do a task like vacuum the carpets. The situation then becomes one not only of computer information-storage capability, but also of the sensing required to do this task. Let's examine what he said. In the first place it probably will be necessary to place into the storage element of the computer a complete track of what the computer must "see" as it moves around. The predetermined track will then serve as a reference. We know there are guided missiles able to fly across continents by using a prepared radar or photomap of the terrain beneath it. It simply flies such that the picture it is taking *as it is moving* must match the picture which has been programmed into it earlier. It steers itself so this will be true. Think about that for a moment! A cup of coffee (or something) helps while thinking, incidentally.

THE ROBOT'S VISION

Now, the previous discussion simply means that the mobile system—call it what you will, guided missile or robot—is supposed to move at some *planned speed*; therefore, the frame advance of the programmed path will continue to present a changing picture *at a proper rate* so that it should be matched by the picture seen by the object in motion as it steers itself.

In a robot, however, it might be that a variable movement speed will be desired. That means that it must have a very, very fast way to search in its memory banks for a picture to match what it is instantly "seeing." This will be a little different from the constant-motion, constant-speed concept associated with the proven technique used in the guided missile.

Our computer expert, Jim, also pointed out something which we had not given a great amount of thinking time to, and that was the probable necessity to have sensors able to measure in three dimentions and advise the robot's computer of its spatial location. We, as humans, do get three-dimentional information, length, height, and width through the use of our marvelous eyes. Our brains work so fast we are not really conscious that when we look at an object we are comparing its position. We see it in relation to a wall behind it, to the floor and the ceiling, and to other objects in the room, so we know exactly *in three dimentional space* where it is located. We also know the shape, related to other shapes, the color, related to other colors, and so forth. We absorb all this at a glance. It makes one feel very humble when you think of just how marvelous this *human* machine of ours really is!

So suppose that we do provide our robot with, say a small, short-range radar, one able to scan space and locate objects with respect to its surroundings. We will talk about radar sensors in a later chapter. Then we can provide our robot with a shape memory, not of every shape possible, but probably the basic shapes things fall into. This would be great research for artists. They would probably tell us that everything imaginable falls into shapes like a cone, box, rectangle, circle or sphere and so forth. Thus the robot can search its memory for basic shape identification—not exactly as we do, but good enough to make its own identification, of objects. Just how many kinds of 'basic' shapes are required would have to be determined.

Now there is that mobility problem again. It seems that with sensors like these the robot will need a measuring device which can tell it how far it has gone and in what direction from some starting point in the house or yard. This does not seem to be a big problem. We program a given path into the robot by, in effect, telling the robot, "Go 7½ feet forward, then execute a 90 degree right turn, then go 3 feet forward, then execute a 20 degree turn left—and so on. Since our robot probably will move on wheels, it is a simple task, using currently available technology, to equip one wheel with a distance measuring device called a transducer. This will give, say, a voltage or series of pulses related exactly to the rotation of the wheel.

We remember that the distance around the periphery of a wheel is a fixed distance, so the number of revolutions times this distance will be the distance the robot moves. Another transducer, converting degrees to pulses, will tell how much some linkage lever turned the steering wheel and in which direction it is turned. So the actual recording of where the robot goes is probably not *that* difficult. When this information is compared to the basic program put into it—which says, "four feet straight, then 20 degrees turn right"—the robot may be able to execute this without real difficulty.

There are drive mechanisms which can be used to give *variable* gear ratios, thus variable speeds and the line, for physical movement of the robot's limbs and wheels. We will need to understand something more about gearing. You don't just use them to get more power from a given rotating device such as an electric motor. That is forthcoming in a page or two.

ROBOT TOYS

The future will bring us many new robot-related toys. Already they are making their appearance. Even many years from the time of this writing newer toys with more-complex computer controls and mechanical assemblies will be coming to the forefront. How else can our children and their children become familiar with the big robots they will have in households and see at work as years move swiftly by.

It reminds us of a toy automobile just developed. It steers and moves by voice commands, probably like Elektro operates. Now that is not so remarkable. We can show how this can be done quite easily with currently available devices. What *is* remarkable is that the manufacturers plan to produce a *million* of these units. People do want robots, if only for the amusement of their children and themselves.

How would it work? The operation is probably sequential. A sound, of the proper frequency, "heard" by the toy car via its microphone may cause a stepping relay unit to operate in sequence. The first sound could order it to "Go left." Then a slight pause in sound transmission followed by a second sound may command the car to "Go right." A third sound could say to "Go straight," and a fourth sound could order it to "Stop." This sequence would repeat over and over with the four sounds. So if you wanted the car to start off you say "go" and to make it go straight you say "Go straight," with a slight pause between words. These would be voice commands.

Yes, you say, television sets already have some sound-operated, channel-changing systems called "Space Commands." These operate in the audio range above human hearing. They cause a stepper motor in the TV (or an electronic stepping transistor circuit) to shift from one channel to another. These units may step through all channels automatically if you hold the sound button pressed down, or they may move in step from one channel to another as you press and release the sound initiating button in impulses. We agree that the use of sound commands to machines is already here and operating.

But what is interesting about the possibility of voice commands to a robot is this: Everyone would like to be able to give the robot voice commands and not have to use any devices like a hand-held calculator to program such commands. Many people fear they would not be able to understand a programming operation. They might make a mistake and give the robot a wrong command. The voice command system, they imagine, would be more simple and direct, and everyone is certain that the robot will understand what it hears and not make any mistakes in what it is supposed to do. But humans sometimes do not understand verbal (or written) information either. Mistakes sometimes result even when using verbal communication between homo sapiens highly skilled in this form of communication. And, of course, a robot isn't as smart as we are—today.

There could be problems with voice communication to a robot. We should agree in the beginning that not everyone should be able to tell the robot what to do. If everybody could command the robot, it would be like a home with each child and parents having equal authority to issue all kinds of commands. In the case of the robot, imagine the confusion if everyone gave it commands and some were contradictory. It reminds us of a Star Trek program we once viewed wherein the villian was a wandering space robot with the ability to destroy everything that was not perfect. It had to destroy itself because Captain Kirk persuaded the villian that it, itself, was not perfect. Would our home robots go up in smoke if everyone in the family told it what to do. Or would it, like a lovely lady robot seen in a film, just respond in a tired, puzzled voice, "It doesn't compute."

Mother and father, perhaps even an older child, might be given the authority to voice command the home robot. A yard robot might be restricted so that it would not accept commands from anyone but the parents. It could be more dangerous if confused, as it probably would have the ability to do much more difficult work and thus have a greater strength and speed than the house robot. Certainly, you wouldn't want everyone, —friends and neighbors, and even criminals who might break into the home—to be able to tell *your* robot what to do and what not to do. Can't you imagine the frustration of a home owner if some thieves were able to break in and tell the robot, "Hey. You. Help me load this stuff."

VOICE IDENTIFICATION

Voice identification is a field which has progressed beyond the possibility stage. In fact there have been many scholarly articles written which tend to show that voice frequencies may serve a better means of individual identification than finger prints. Of course much electronic equipment is necessary to analyze a voiceprint as contrasted to a finger print. And it is the possible error in the complex equipment which may make this method less preferred than the fingerprint method. But it is possible and feasible. With this system it would be possible to make your home robot(s) respond *only* to those voices which have the authority to give it commands.

It will take a bit of analysis and psychology studies to determine who, when, and how to voice program a robot. We can think of some gray areas. If the robot is to teach children, or be companion, or monitor children, then should they be allowed to give it some limited commands and have it properly respond?

For example, suppose that the teaching session has gone on for quite a while and the children are tired and want to play. Should they then be able to command the robot, "Let's play some games," and even specify the kind of game? The robot, having previously received commands from the parent as to what kinds of games are permissible under the present situation, may respond by saying "Yes," or it may respond by saying "We are not to play that game today, choose another."

Then also, what does the robot do if the children decide to go ahead and play the forbidden game anyway? The robot might make a very loud noise alarming and warning the parent, or it might take some other action which you might imagine to be more suitable than we can at this moment.

In any event deciding who programs the robot by voice command and who cannot will be a problem to be solved. We feel certain that much testing will be involved before completely satisfactory answers are obtained to this question, before this method of command might be used. But then, such simple commands as "Answer the door, Answer the telephone, Turn on the air conditioner" and so forth might be voice commands which would not pose a problem, even if everyone gave them. You will probably think over this situation and come up with some good situations and problems of your own. The solutions must be applicable to your own family life style and situations. You can have a ball discussing these ideas with your friends and neighbors.

The more complicated a robot becomes, the more things it can do, the easier it is to program, and the more reliable it becomes, the closer we approach the android. This, as we have indicated, would be human-like in many aspects, and some believe that it will always be humanoid in appearance, even to the extent of performing the difficult function of walking.

EMOTION AND LOGIC

Let's talk for a moment about emotion and logic. The one thing which a robot, android, or similar machine *won't* have is emotion. It is a logical machine, we all know that. He (she or it) is *all* logic—logic from those millions of tiny solid-state devices, resistors, and capacitors making up the central processing unit.

The central processing unit is that part of the machine which does what it says—it processes (acts on, analyzes, separates, computes, sends out, stores, etc) all information it receives from whatever sources. It has no emotion, and this is what makes it basically different from people. If it is told to grasp something, like a person's arm, it can do so, and its logic and *pressure feedback* will tell it not to exert too much pressure. But the concept of being gentle, or of giving pain if too much squeeze is used, or perhaps breaking something if it grasps too strongly, just cannot be a part of its make-up. It cannot understand this kind of situation really. It might be programmed such that if its eye sensors detect a change in your facial expression such as the compression of your lips, shutting of your eyes, a change of your color from red to pale, or something like that, it would know harm is being done by grasping your arm too firmly. But that wouldn't be logical, would it?

In our next chapter we will begin our somewhat technical discussion of these machines, as to their *motion* capabilities and how they are accomplished. Later we will examine some power sources and study some information regarding these motion power sources. We want to learn all we can about everything that is needed to design and make a good, reliable, intelligent robot, don't we?

CHAPTER 2
GIVING A ROBOT MOBILITY

We already mentioned the problem of making a robot "walk." Studying the way *we* walk reveals to us that humans constantly unbalance themselves, then thrust out first one and then the other foot, forward, backward or even sideways, to regain balance. We also indicated in the first chapter that we want our robot to be as physically stable as possible, which simply means that it should never become unbalanced. Practically though, we know there will be extremes to cause some unbalance in even the most stable machine. Even your automobile can turn over under certain conditions.

A WALKING ROBOT

Although the problem of having a walking robot is very difficult, especially if we must meet demanding stability standards, it is, nevertheless, possible to some extent. For example look at the little machine of Figs. 2-1 and 2-2. Here the machine moves under spring power, advancing first one leg and then the other, and if the terrain upon which it operates is a slight downgrade, then it does seem to walk forward in the manner that we do. The incline gives it the slight unbalance necessary to make it advance.

Notice the arrangement of the leg and foot linkage shown in Fig. 2-3. The shaft is shaped so when it rotates it causes first one leg to lift and move forward and then the other.

Fig. 2-1. A spring-powered walking machine.

The spring power provides constant torque by means of a tiny but effective centrifugal governor at the top inside the mechanism.

The shape of feet are important. The inside "spurs" keep the

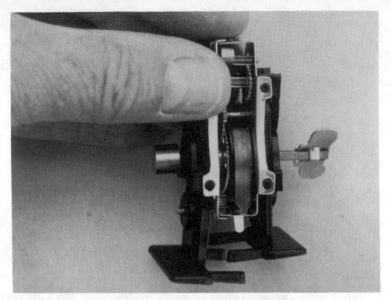

Fig. 2-2. The spring, legs, and feet of the walking machine.

machine balanced on one foot as the other lifts and moves forward. Figure 2-4 shows how each foot and spur gives the machine a fine balance. The spur arrangement on the foot is necessary so that the feet can rotate inside each other.

From our illustration here we see that it can be done. Perhaps you will want to experiment more with this arrangement to have a real walking robot. This model is available from the Gallery in

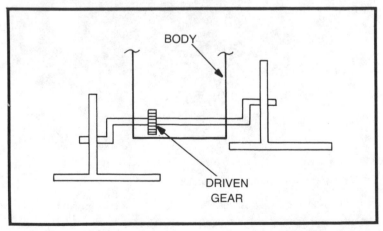

Fig. 2-3. This shows the basic arrangement for making the feet and legs move.

Amsterdam, N.Y. 12010. We did find, however, that its balance was somewhat precarious, even with the wide feet and spurs. It would fall over with even slight irregularities in its supporting surface.

Walking toys have been on the market for years, and there will be more. Some use a quick foot-shuffle movement to cause the walking effect. You may want to visit a toy store and examine some of these to see how they accomplish the walking movement.

Remember, the lower the center of mass the more stable the machine will be. And the wider the support base the more stable it will be. So you'll have to think of these in relation to the esthetics of the robot's appearance.

A RUSSIAN DEVELOPMENT

The Russians performed many studies about walking machines to enable them to explore the surface of other planets. There is

Fig. 2-4. Bottom view of the walking machine showing the feet and inside spurs.

some feeling that walking robots may be able to traverse unknown terrain better than a wheeled vehicle.

One Russian study described a six-legged walking spider with a four-eyed manipulator arm. The parts used in this development were off-the-shelf items from the aerospace industry. It was said that the spider, because of its *six* walking legs, could move across almost any terrain by varying its gait and posture. The fastest movement resulted when only three legs were used. This produced a forward motion of 6 km per hour. It was indicated that this spider could stand, crouch, and even climb up steep slopes, with special attachments on its legs.

The spider's laser eyes were said to be able to measure distances quite accurately down to an error of 5 mm. The four-eyed manipulator actually was an arm with four photocells. It could discern an object then resolve in which direction the arm had to be moved by a hydraulic power source to grasp the object with its tactile sensor. By the way, a *tactile* sensor is one which gives a sense of touch. So presumably, the spider could "feel," through proper feedback, the object it grasped. Yes we might certainly see "commercial" walking robots, but they might scare the dickens out of us. And they might not be a two-legged variety either.

POWER SOURCES

We must get back to the basic subject of this chapter, and that is power sources. We have seen that spring power is one means of obtaining locomotion. Now let's consider the wheeled robot. This, as you would expect, will either use electric motors with gears, clutches, and speed control devices, or some other system. This might be pneumatic (a "motor" which runs on air pressure), or hydraulic (there is a "motor" which runs on oil pressure), or possibly even rocket power in some applications. This latter probably would only be used in space, and (don't grin) not in the home.

TRIPED POWER

We begin by thinking of three wheels (triped), two fixed on an axle and one affixed to a shaft which can be turned for steering and possibly also powered to give motion. This might resemble Fig. 2-5.

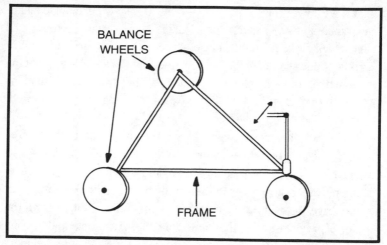

Fig. 2-5. One concept of a triped base.

This arrangement will give the robot stability if the size of the triangle is large enough so that the center of mass is equidistant from each side and low.

The next question to arise is which wheel or wheels to power to give the frame movement. By the way, the frame actually could be round and have just three wheels. This would make it easier to use a cone-shaped body, as Klatu had in Chapter 1. Also we must consider that a four-wheeled arrangement might be used for the robot base and driven as any small four-wheeled toy or car is driven and steered. This would have excellent stability, more than the three-wheeled arrangement. But the problem of having it move around exactly where we might want it to go could be more difficult, requiring additional steering linkages and possibly more movement of the wheels. Since the three-wheeled arrangement is the simplest, for experimentation at least, and does seem to answer our basic requirement, let's concentrate on that a little further.

There are three reasonable possibilities for motion. First we could power the steering wheel so that it pulls or pushes the frame. Second, we could power one rear wheel so it pulls or pushes the frame. Third we could power both rear wheels through a differential gearing arrangement like your automobile has. This allows one wheel to rotate faster and farther than the other when the frame is turning.

If we power the steering wheel, the steering mechanism, shaft, and attachments will have to be capable of supporting the motor and gear weight, which must be close to the wheel and turned with it. If we power one rear wheel we will not need a differential—if we let the other wheel rotate freely on the axle, but there might be situations where the unbalanced drive could cause steering or mobility problems.

We can imagine an obstacle located in the path of the free wheel, causing the wheel to drag slightly. Then the robot frame would tend to turn in that direction, and steering would have to compensate for the uneven motion of the frame. With this single rear wheel drive the point of push or pull is off center, not through the center of gravity (c.g.) of the frame. The use of a differential—giving drive to both rear wheels and yet permitting one to rotate independent of the other, as required in a turn—makes the frame move like your car moves. That is, steady, giving proper drive parallel to the center of gravity. The problem is, of course, that a differential gear is required. If the right size is available and if you like this method, then use it. The advantages are many. The steering wheel only steers; the thrust is in the proper direction; the drive can be accomplished with a single motor. Then too, forward, reverse, and turns are properly under power. The only disadvantage might be the cost and availability of a differential unit.

Motor Power Required

Once the general system for propelling the frame has been determined, the next problem is determining what size motor must be used and what gearing must be provided. These are questions basic to robot design. They apply to the movement of other parts of the body such as the head and arms.

There are two basic methods for determining the motor horsepower and required gearing, if any, to properly propel the robot. To begin our considerations we first need to have some close idea as to the total weight of the robot. We can estimate this by adding the weights of the various items we plan to put into the machine such as the following:

- The battery.
- The skirt or outside cover.

- The metal parts of head, arms, and frame.
- The electronics package.
- The motor.
- The gears, etc.

You see, we try to estimate the weight of each of these things and then add them all together to get the total weight. Of course, we could just say, "Well, this robot will not weigh over, say, 200 pounds and the motive system will be based on that weight." If we do this and the robot is lighter, then our propelling system might be larger than needed. That doesn't cause any problems. It just means the robot might go faster than we planned. And since we must control the speed of its movement, that is if we want a really good robot, having the additional power to move faster simply means the system works at less than maximum capability. That is all right. It isn't the most economical way to go, but then, if you have a few bucks to spare, well spend them. You might want to add things to your robot later.

Now, the first of the two ways to determine the power required is the experimental way. You take the frame, put the wheels in place, look at the kind of terrain over which it will operate, and then select some motor and see if it will drive the frame. Try it out. If the motor stalls and won't make it go, you then proceed by adding gears—first you try two, then if larger gearing is required, you add another to make a three gear arrangement. We will take up some consequences of using many gears just a little further on. You might want to glance ahead and look that over then come back and re-read this section.

Now, to evaluate the motor and gearing you have experimentally selected, see if the choice moves the frame at the required speed when it is loaded to approximate the final weight of the robot. Also see if it has the required ease of motion you need.

It is well at this point to determine if the battery drain is excessive. This would cause the battery to run down quickly. Also determine if the motor rotational speed is fast enough so that the drain is light. That way you get the longest possible battery life per charge. We assume a rechargeable wet cell of some kind powers this motor. Notice in Fig. 2-6 the batteries, motor, and differential gear of an electric car.

Just to have the robot move isn't enough. The motor may turn so slowly that it will use too much current. Of course if it is slowed by a controller resistance, that will limit the current, but we are talking about the approach to stall speed of a motor. Adding some gears will permit the motor to turn faster than a direct drive but will make the robot move slower. You *can* experiment until obtaining the right combination of motor speed, battery drain, and robot movement. But trying different size motors and gears using your base platform loaded with the estimated weight can be time consuming.

Selecting Motors

Another fact about your motors. In a robot you will probably be using dc motors. They run directly from a battery like the blower motor, or the seat-adjusting motor or the window up-down motors in your car. Note, however, that there is available some conversion equipment which makes it possible to run AC motors from batteries. But the start-stop procedure may be easier with the DC motor.

Another thing. Motors vary in size and shape for a given horsepower. Actually you want a motor which has a small diameter armature. Length is not so important. These can be started and stopped quickly. It should have low inertia. Those with big, heavy armatures will coast when cut off and will take too long to reach running speed. These large-armature motors also draw a lot of current as they build up to speed. The motor we need is classified as a servo motor. There are many types and sizes. Some have built-in gear trains—such as the seat adjusting motors and the window raising motors in your car—which may give you exactly the torque and speed you desire for whatever part of the robot you may want to have move. But, if you know the horsepower motor you need and the gearing ratio you want, you simply can go to a manufacturer's catalog, or write to a motor manufacturer and ask them for what you need. You might even ask them for a recommendation and tell them the weight of what you plan to move and the speed at which you plan to move it.

Again we come back to the AC motor. It may turn out that you just cannot find a DC motor to do your job. Be aware, then, that you *can* get an AC motor. But you'll need to buy a converter to change the DC from the battery into AC for the motor. In this case the whole system to control and adjust the speed and direciton of rotation of the

Fig. 2-6. The power system of an electric car.

robot's propulsion motor will probably be AC operated. There are many AC servo system controllers.

adjust the speed and direction of rotation of the robot's propulsion motor will probably be ac operated. There are many ac servo system controllers.

Does the experimental approach seem complicated and expensive? Yes, it could be. You start with some motor size, then you adjust and change, perhaps having to buy several motors before you find the one to make your robot move as you desire (or move individual parts of the robot, such as the arms). It may take quite a bit of money. It would be better if you could just calculate, at least roughly, the size motor you'd need and start from there. That would be cheaper and more direct, and you'd have less worry and trouble getting the drive system organized and laid out. It *is* possible to do this, even if a small amount of mathematics (Wizard symbology—and how they love it) is required. So don't give up. We will make some assumptions to keep these calculations (Wizard magic) as simple as possible.

Of course, once again, do not rule out the possibility of getting some help from a motor manufacturer. If you know the weight of your robot or the weight it must lift with its arms, you're well on the way to specifying the motor requirement. You'll also need to specify the speed at which you plan to have it move, its battery source voltage, the wheel size and layout, and the directions (forward and reverse). You might write all these things down and send them in a letter to an electric motor manufacturer such as Westinghouse, G.E., or some other servo motor manufacturing company. Ask for their recommendations for the motor(s) and gear train(s) which you will need. They can also give advice on controlling the motor speed and direction of rotation. These manufacturers are most helpful, if slow, and do not mind such requests.

MOTOR AND BODY MOVEMENT

Let's now consider some engineering (what a ghastly term) concepts related to the design of a robot mobile system. We want to try to determine the horsepower and speed required of a motor able to move a robot. We will assume a total robot weight of 100 pounds—that's almost human and the figure is a nice one to work with. Let's assume a stop speed of 20 miles per hour, and we want the robot to be able to reach that speed in, say, about 10 seconds. Of course it would be easier to assume a constant operating speed, without adjustment, in the range of, say, 10 to 20 miles per hour. We could simply have the motor on or off and either FORWARD or REVERSE. But that doesn't make it as flexible or as usable as we'd like. It should be able to go toward a person or object, moving fast at first, then slower and slower as it gets close. That requires speed control. We can add the feature later, of course. But to get back to motor size, we must also make an assumption as to the operating terrain of the robot. Will it always be on a level floor? What kind of floor will it be? Rugs may give more friction (traction) than a stone, marble, hardwood, or concrete floor.

LEVEL MOTION

It is said in mechanics textbooks that if you have an object of *any* weight resting on wheels which have no give, and they are on a perfectly level floor which has no give, then there is no friction between the wheel and the floor. With these conditions the *slightest* push perpendicular to the weight will cause the body to move. That is, the wheel will roll (or slide). Figure 2-7 illustrates a wheel which gives on a floor which gives, the true situation.

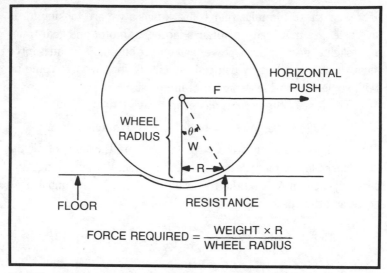

Fig. 2-7. A wheel on a floor that gives has resistance to motion.

But in our frictionless example under discussion we must also assume that there is no friction on the axle of the wheel. In actual practice, however, the wheel will probably be rubber tired, the floor or surface to be crossed may be "soft," so there will be compression. Thus there will be some force (R) trying to resist rotation or motion. This does not include any resistance force due to incline, roughness of surface, wind resistance and so forth, which may or may not be present.

A BASIC FORMULA

What we apparently need, then, is some formula which will give us an idea (a ballpark figure) as to the approximate size of the powering motor we will need for a given *weight* robot. We will assume rubber tires and slight compression, like an automobile tire has, so that some friction will be taken into account. We will want to have some means of making the motor size determination for a level surface or for a slight incline, whichever may be the worst situation the robot may encounter. Because if we assume the worst case, we know the motor system will work all right in any other situation of less difficulty.

There are two formulas which some physics texts use to determine *average horsepower* needs for a body in motion, such as this.

If we use these formulas that ballpark figure we are looking for is available. We may want to obtain a motor, or motor and gear train, for a slightly greater horsepower and torque output. Torque is force times lever arm. But by using these calculations we can come up with some reasonable estimate of motor size.

The first of these formulas simply says that

Average horsepower = total force × average velocity

We need to define the terms. Total force is the sum of all the unbalanced forces plus the resistance to motion. Without going into a long explanation we find the following to represent the unbalanced force and the resistance:

- Unbalanced force = $\dfrac{\text{weight}}{\text{gravity}}$ × acceleration (a)

- Resistance = weight × sin of angle (θ)

This is essentially the old mechanics formula $F = ma$, and the weight (W). The angle (θ) is the angle of the incline the robot must climb.

Let's now consider an example problem for a robot of weight (W) of 100 pounds, gravity equal to 32.2 ft/sec^2, and an incline angle of 6 degrees—which is probably good enough for any "level" flooring. Writing the equation in a form which can be solved by our calculator we can use it to do the "pick-and-shovel work" of computing.

First, however, we will need to get some values such as acceleration (a). To find this value, we need to know the distance in which our robot will build up its speed from zero to 20 mph, which we set as the top speed. Let us say that this will require a distance of 100 feet. The acceleration is then computed:

$$\text{Acceleration} = \frac{\text{final velocity} - \text{initial velocity}}{\text{time in seconds}}$$

The velocity must be in feet per second.

Note also that we can find the time from:

$$t = \frac{\text{distance in feet}}{\text{average velocity}} = \frac{\text{distance in feet}}{\text{final velocity}_2 + \text{initial velocity}}$$

Note that the *average* velocity is initial *plus* final velocities divided by 2.

Velocity is the speed in feet per second; therefore we need to reduce the 20 mph to feet per second. This is done by stating

$$\text{Miles/second} = \frac{20 \; \dfrac{\text{Miles}}{\text{hour}}}{3600 \; \dfrac{\text{sec}}{\text{hour}}} = 0.00555 \; \text{mps}$$

$$\text{Ft/sec} = (0.00555) \times (5280 \; \text{ft/mile})$$

$$\text{Final velocity} = 29.33 \; \text{ft/sec}$$

Thus the average velocity is $29.33/2 = 14.6555$ ft/sec, as our initial velocity is zero.

We need next to determine the time (t) to get to this final velocity. Recalling that we want our robot to reach this speed in a distance of 100 feet, and recalling that the initial velocity was zero, we use the following:

$$\text{Time} = \frac{\text{distance (ft)}}{\text{average velocity (ft/sec)}}$$

$$= \frac{100 \; \text{ft}}{14.66 \; \text{ft/sec}}$$

$$= 6.82 \; \text{sec, or rounding,}$$

we would state the time to be 7 seconds to make further calculations easier.

With the previous values for time we now calculate the acceleration from:

$$\text{Acceleration} = \frac{\text{average velocity}}{\text{time}} = \frac{14.66 \; \text{ft/sec}}{7 \; \text{sec}}$$

$$= 2.09 \; \text{ft/sec}^2$$

If we now remember that one horsepower (hp) is equal to 550 ft-lb per sec, we can calculate the horsepower needed by our motor. We use the equation

$$\text{Horsepower} = \frac{\text{force} \times \text{average velocity}}{550}$$

We need the value for the force (F) in pounds due to weight and to resistance to motion or friction. A formula which we can use says

$$\text{Force} = \frac{\text{weight} \times \text{acceleration} +}{\text{gravity}}$$

$$\text{weight} \times \text{sine of the incline angle}$$

or

$$F = \frac{Wa}{g} + W \sin \theta$$

We said that the incline angle our robot could be expected to encounter in this example was 6 degrees. If you plan to have your robot climb up a driveway entrance which might have a larger angle, say 30 degrees, then the sine of that angle should be used.
Let us now calculate the force F.

$$F = \frac{100 \text{ lb}}{32.2 \text{ ft/sec}^2} \times 2.09 \text{ ft/sec}^2 + 100 \text{ lb} \times 0.105$$

F = 6.49 + 10.5 = 16.99 pounds, or rounding again, 17 pounds.

To find the horsepower we must divide this by 550 ft-lb per sec, which is equal to one horsepower, and then multiply the result by the average velocity (14.66 ft/sec).

$$\text{Horsepower needed} = \frac{17 \text{ lb} \times 14.66 \text{ ft/sec}}{\dfrac{550 \text{ ft-lb}}{\text{sec}}} = \frac{249}{550}$$

$$= 0.453 \text{ hp}$$

And this is what is indicated as the ballpark figure to move our robot as we have indicated. Remember, the weight was 100 pounds. A larger weight or faster movement, or shorter distance to reach the maximum speed will all change the value of horsepower we came up with. Also, the steepness of the incline, if different from our figure of 6 degrees, will cause some change in value.

What is the size of a 0.453, or roughly a ½, horsepower motor? As a gasoline unit it is not very large. Most lawnmowers use 2 to 3 horsepower engines. As an electric motor it is about the size of your air conditioning blower system motor. They usually use from 1/3 to ½ hp motors.

Now we begin to consider some other factors. A direct drive system—without gears or pulleys or the like—will usually require a lot of battery energy. So we will now probably want to consider the use of gearing to get the required output. This can also reduce the size of our motor so that we can save on weight, cost, and current drain. This will require a trade off. We must sacrifice something of our initial specifications, and the easiest one would be the speed of robot movement. But we want our batteries to be as small as possible and still give a good operating life per charge. Cost, weight, size, and perhaps other factors are considerations at this point.

A SECOND FORMULA

Since we mentioned two formulas, we will give you the second one without elaborating on the explanation. Klatu indicates that too much magician symbology might become boring even to him. So, since both formulas give the same results, we include this one for those of you experimenters who might want another go at the method. In this formula we consider the kinetic energy (KE) and the potential energy (PE) demanded of the system. The potential energy comes from the height change due to the angle of 6 degrees. The KE term is the same as before. We are to consider *changes* in these values from zero to terminal speed.

Again we need the initial velocity and final velocity and the distance to reach final velocity. As before we find the *average velocity*, and time.

$$\text{Change in KE} = \frac{W}{2g} \times V_f^2 - V_o^2$$

where

V_f = final velocity
V_o = initial velocity
W = weight
g = gravity

The change in PE = WH = W × s × sin angle, where s = distance.

From these symbols and the equations we can calculate the total energy change which is equal to the sum of KE and PE.

Change in energy $= \left[\dfrac{W}{2g} (V_t^2 - V_o^2) \right] + (W \times s \times \sin \theta)$

And with the appropriate values plugged into the above equation we get

$$\text{Change in energy} = \frac{100 \text{ lb}}{(2) \dfrac{32.2 \text{ ft}}{\sec^2}} \times (29.33 - \text{zero}) \; \frac{\text{ft}^2}{\sec^2}$$

$$+ \; 100 \text{ lb} \times 100 \text{ ft} \times 0.105$$

This works out to be

Change in energy = 861.55 lb-ft-sec plus 1050 lb-ft

$= 1911.55$ lb-ft (motor torque required)

Now, this figure must be divided by the time which, remember, is distance/average velocity. We come up with the same time as before, 7 seconds, with exactly the same method of calculation. Finally, then, we solve for horsepower, which is the change in energy divided by the product of time and 550 ft-lb per second (one horsepower). We get:

$$\text{hp} = \frac{1911.1}{7 \times 550}$$

$= 0.4965$, or rounding 0.5 hp

If you are careful making your calculations you will find both values of required horsepower agree exactly. Our difference is due to rounding values somewhat. But both answers are close enough for what we wanted originially—a ball park figure.

GEARS, GEAR RATIOS, AND PULLEYS

We come now to the second basic part of the mobility situation. We can use gears, Fig. 2-8, or pulleys to build up the torque to the drive wheel. This allows the use of a lesser torque motor, but remember that when we increase the torque with gears, we reduce the speed of the output shaft too.

Figure 2-9 shows the use of spur gears in a toy robot. Spur gears are also used in the small DC motors shown in Fig. 2-10. A

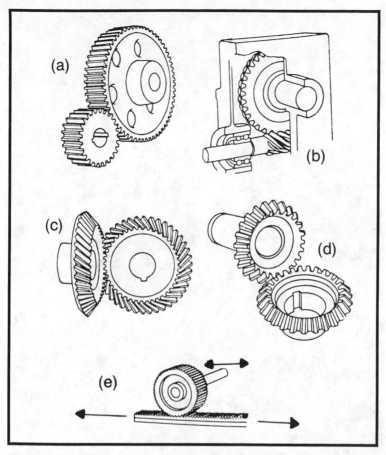

Fig. 2-8. Typical gears. (a) Spur gear. If the small gear drives the big gear, torque is increased but speed reduced. (b) Worm gear. Provides great torque but very slow speed. (c) Spiral bevel gear. (d) Bevel gear. This and (c) are often used in differentials. (e) Rack and pinion. These are commonly used to convert rotary motion to linear motion. Either of the two can be made to move.

motor such as this could be used in the arms of a robot. The pulley system in Fig. 2-11 gives some advantage over gears as it is simpler, perhaps, but will require a larger physical space.

Now recalling that with the use of gears in the manner planned we will reduce the speed of the output wheel. Let's do some calculations. Assume that the electric motor will run at about 7000 rpm at full speed. Manufacturers publish specifications on their motors to give the running speed versus torque at various combinations. They give what is called a torque-speed curve for each motor,

Fig. 2-9. Plastic spur gears in a toy robot.

an example of which is shown in Fig. 2-12. Now, in order to determine our gearing ratio we begin by considering how fast a robot base
platform drive wheel of, say, a diameter of 10 inches must rotate to
travel a distance of 100 feet in 7 seconds. We recall the circle formula
which gives us the distance around the wheel as;

$$\text{Circumference} = 2\pi r$$

Fig. 2-10. This small 12 V DC permanent-magnet motor with built in spur gears could be used in the arms of a robot. Note the golf ball.

where

$$2\pi = 6.28 \text{ a dimentionless number}$$
$$r = \text{wheel radius in inches (5)}$$

Thus the distance around the circumference is 31.40 inches. The wheel will turn one revolution going this distance over the ground.

Fig. 2-11. A pulley drive system driven from a geared motor.

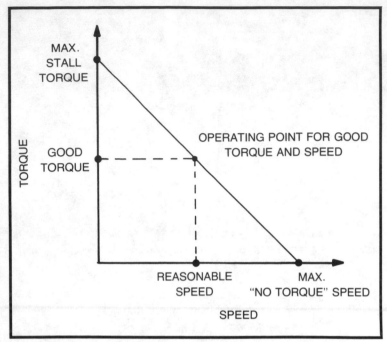

Fig. 2-12. A typical torque-speed curve for a motor.

Next we recall that the final velocity our robot is to have is to be 20 mph, or 30 feet per second (rounded off slightly). Thus the wheel must turn

$$\text{Revolutions per second} = \frac{30 \text{ ft/sec}}{2.62 \text{ ft/rev}} = 11.45 \text{ rev/sec}$$

and in one minute must turn 60 times that, or 687 rpm. That will be the final speed of our rotating drive system. If, then, we use a gear reduction to reduce the motor speed from its 7000 rpm down to 687 rpm—agan let's round off, making the final wheel speed 700 rpm for simplicity—we see that a 10-to-1 speed reduction gear will be needed. The rounding off here will give our robot a slightly faster movement, but not enough to make an difference in the overall operation.

What will the 10 to 1 gear reduction permit us to use in the way of motor torque? It will be less than the direct-drive torque calculated earlier because the reduction gearing used actually multiplies the effective torque from the motor, in this case by a factor of 10. In the case of a pulley, this is the ratio of the output pulley radius over the input (motor) pulley radius. Thus we could consider a belt and pulley arrangement as in Fig. 2-11, which would eliminate calculating teeth in a gear ratio arrangement. But remember that the *output* torque will be the input (motor) torque *multiplied* by the gear ratio. The speed of the output shaft will be the speed of the input shaft *divided* by the gear ratio. The gear ratio is usually symbolized by n.

So we reduce the output shaft speed to 1/10 the input shaft speed, and the *motor* torque required to produce the required torque at the output shaft (axle) will be reduced. You can use a smaller motor and its current drain should be much less. By the way, you can find gearing ratios to meet almost any combination of needs as to speed and torque at machine shops, electric shops, or from gear manufacturer's catalogs. You can use belts and pulleys too. These can be found in hardware stores, air conditioning repair and maintenance shops, and so forth.

You may find that using a pulley drive is easier than a gear-drive system. It does permit some flexibility in mounting the motor with respect to the output shaft (axle). It is also possible to obtain a centrifugal clutch which will engage when the motor speed reaches a given value, so that when the motor operates below this value, there will be no drive to the wheels. This may be very useful if you use a gasoline motor because with that type motor you do not want it to stop running, although you may not want the robot which is powered by it, to move. The clutch, which disengages at low speeds, accomplishes this latter purpose. With the electric motor drive system you simply turn it off (or the robot does) if you or it doesn't want the robot to move.

CHECKING THE EQUATIONS AGAINST KNOWN ELECTRIC CARTS

In order to make a check to see if the equations would hold up, at least in the ballpark area, when used against some actual electric motorized units, we made a check of some golf cart manufacturers. We asked to see what size motors they used and to obtain the other data necessary to fill in our first equation. We found from the first of two manufacturers that they made golf carts using a direct drive (to differential gear) of an electric motor with a rating of 5 horsepower. The unit used a 36-volt battery, had an output-wheel speed of 800 rpm and used an 8-inch diameter wheel. The second manufacturer's unit used a 3½ horsepower motor and a 13 to 1 gear reduction to differential gear. It had an 800 rpm output-wheel speed, used a 36-volt battery and 8-inch wheels. Both carts were designed to carry two passengers and a load of 500 pounds, which represented the machinery, frame, batteries, and such things.

In order to use our first equation we made some general calculations, which gave a top speed of 12 mph (verified by the manufacturer) for the machines. We estimated (based on our last golfing experience) that the machine reached this speed in 5 seconds. We also estimated that the output speed could be obtained on a 6 degree incline. We noted that although the golf cart does climb up a 30 degree slope, it slows down doing this. And then we estimated that two men weighing 250 pounds each were carried in the machine. Hey now! We don't weigh that much! That's just an estimated figure. So, using the basic equation as before

Horsepower = total force times average velocity

We found acceleration by calculating speed from zero to 12 mph in 5 seconds to be 3.52 feet/sec^2. The average velocity we state as 17.6 feet per second, which gives a velocity of 12 mph, the speed stated by the manufacturer. The sine of 6 degrees is 0.105. Now, we can solve for the horsepower, remembering that 550 ft-lb/sec is one horsepower. The results of our figures give 93 + 100 as the force (F). The average velocity is 17.6 ft/sec, and thus the horsepower will be

$$\frac{193 \times 17.6}{550} = 6.176 \text{ horsepower.}$$

And while this is a little higher than what was stated as the direct-drive horsepower used by the first manufacturer, it is close enough so that we have some confidence in our method of making ballpark motor calculations. If you consider the gearing ratio of 13 to 1 used by the second manufacturer you may find that the horsepower figure he gave of 3½ hp may be larger than actually needed. He might use a 2-to-1 gear ratio instead. We rationalize this by saying that perhaps there may be some factor we have overlooked, or maybe they make a good 3½ horsepower motor cheap enough so that they feel it worthwhile to give a little more than minimum power capability to his cart.

You have to remember that all manufacturers* will be looking for product reliability and suitable equipment as well as at cost and material availability. If there are lots of 3½ horsepower motors on the market, certainly it is cheaper to use one than to ask a motor manufacturer to make a special motor *near* that horsepower. In robot design it pays to use off-the-shelf items as much as possible so as to keep costs down and to insure availability of the items which are needed. Too often someone makes a robot but has so many special items in it that general development of the unit is not practical, even for the hobbyist.

A POWERED WHEEL CHAIR

As illustrated in Fig. 2-13 one type of wheel chair uses two separate drive motors, one for each large wheel. The controlled ratio of speed between the motors governs whether you go straight or turn. In front of the chair, two pivoted, unpowered wheels, extremely free to turn, immediately line up in the desired direction of motion, while furnishing support for the chair occupant.

The torque from each of the two small battery operated motors, Fig. 2-14, is applied first to a worm gear, Fig. 2-15, and then to a 3-inch pulley attached to the *gear* shaft. This pulley then drives a belt which goes to a larger (11-inch diameter) pulley fastened to the inside of the wheel chair's large wheel. Thus there is a tremendous gear reduction between the motor shaft itself and the output wheel, which has a 20-inch diameter. We recall, however, that the larger the wheel on the *output*, the more torque is required to turn it, in view of the equation that force times distance equals torque. Thus the retarding torque due to friction on a 20-inch wheel is very much greater than that on a smaller wheel of 8-inch diameter.

---

* One source of electric motors suitable for robot applications is Palmer Industries, Box 7071, Endicott, N. Y., 13760. They make Bikepower kits using electric motors.

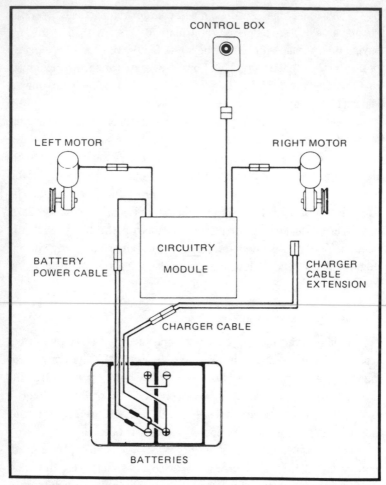

Fig. 2-13. Power drive components of an Everest and Jennings powered wheel chair.

Now, on the motor, specifications such as output speed equal to 320 rpm and motor shaft torque of 15 *inch*-pounds are given. Note that this is different than the *foot* pounds. The inch-pounds reference means a force equal to 15 lb at a *distance of one inch* from the center of the axle is produced by the motor. Note also the slow speed. That there might be an internal gearing was not disclosed, but this is a very slow-speed motor.

That brings up another point concerning electric motors such as these. If the motor is series wound it could have a high speed and low

Fig. 2-14. One of the two motors used to drive a wheel chair. A worm gear fastens to the shaft. Sufficient torque and speed are produced to propel a 250-lb person at 2 to 3 mph.

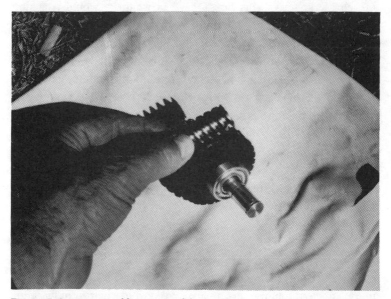

Fig. 2-15. A worm gear. Many turns of the worm are required to produce a single turn of the worm wheel, providing a very large torque.

Fig. 2-16. Bottom view of the joystick control box, showing gimbals and potentiometers.

torque. If the motor is parallel wound it could have a slow speed and high torque. There are also compound-wound motors which have both characteristics. This latter case is what we would consider appropriate for our robot moving mechanism. It is also possible to use a motor with low speed which is of parallel-wound construction.

We could (but we won't) calculate the output speed of the chair by using the ratio of gears and pulleys. If it is done properly we would find the speed to be around 3 to 3½ mph, which is a good fast walk. Normally a person can walk a mile in 15 minutes if he moves right along. That would be 4 mph and is probably about as fast as we would want a home robot to move. Remember, however, that to have the reserve power to climb inclines or ramps means that on level ground the speed would be greater, perhaps even as fast as the 20 mph we had discussed earlier.

Steering the Wheelchair

How is it steered? It is steered by separately controlling the speed of each of the motors. They are controlled by a small joystick steering device mounted on one arm of the wheelchair. A movement of this lever to the left will make the chair turn left, right is right, back is back, and forward is forward. Then the *amount* that you move the lever determines the speed of movement in these directions.

To accomplish this, and to make possible straight forward and backward motion, requires electronic control which can synchronize or vary the speed of the output motors and adjust their torque accordingly so that when you steer forward or backward you will go in the proper direction and not into a partial turn. The control box is shown in Fig. 2-16. Notice the gimbal element with its attached potentiometers. They generate voltages as control signals which go to a solid-state chassis and then to the motors. In a robot we would see these units replaced with the robot's sensors.

Now glance at Fig. 2-17, which shows the electronic module housing the circuit board and transistors which give solid-state control over the motors. The motor speed synchronizing circuits are also housed in this unit. Shown in the figure are the motor control transistors on heat sinks. The circuitry is below them.

It is interesting that we already have available such a drive system which is entirely suitable for a robot. In case you want further information contact Everest and Jennings, 1803 Pontius Avenue, Los Angeles, Calif. 90025. There are other manufacturers

Fig. 2-17. A solid-state motor-control unit.

of course. When looking in the Yellow Pages for wheel chairs don't forget to look for repair places. They stock parts for such chairs. You may be able to obtain parts from them, or even a used chair to experiment with.

The Battery

The battery may be a single 12-volt unit or two smaller 12-volt units connected to give 24 volts. Or it might be a single 24-volt battery. When using battery-operated motors remember that the higher the voltage the smaller the current required from the battery, all other factors remaining equal. So the second battery will double capacity, per charge. But you must, of course, use the voltage the motor manufacturer says to use on his motor. We are just saying that if you have a choice of a 12-volt motor and a 24-volt motor, all other factors being equal, use the higher voltage one. Notice the battery in Fig. 2-18 and its mounting. Note also the pulleys on the large 20-inch drive wheels.

Of course, for your robot, you might just want to power the drive wheel with the worm-gear arrangement and not use pulleys at all.

All batteries used for robots, wheel chairs, golf carts, and electric cars must be rechargeable. Some batteries are specifically designed for these uses. You will need a good charger. When properly used it will keep the batteries in top condition. We have learned that with normal usage a wheel-chair *might* go two days without recharge. But, it is more customary to charge every night to keep the batteries in top condition. This could mean that batteries used in this application (and in robots) should be a special type able to stand up to the very frequent charging and discharging cycle. Think about it.

A final note on the motor. It runs on 24 volts and is said to have a current drain of 3 amperes normally. This means we can find the wattage used by multiplying the two numbers (3 × 24) together and that gives us 72 watts. Since we also know that one horsepower is equal to 746 watts, we can find the motor horsepower to be

$$hp = 72/746 = 0.10 \text{ hp}$$

This is a small motor. But remember, we must always find the

Fig. 2-18. The battery used to drive one model of wheelchair. Note the mounting method.

horsepower required of our output drive wheel on our robot and then find an appropriate gear ratio which will multiply the motor horsepower up to the required output value. The speed of rotation of the output wheel is also critical, so this requires a specification for the motor speed.

WORM GEARING

Worm gearing torque and horsepower calculations are not simple. With spur gears we simply use the ratio of teeth of the input gear to the teeth on the output gear, and with pulleys we use the ratio of pulley diameters to obtain speed and torque. But the worm gear works like a screw, wherein the "worm" rides on the edge of the gear and "screws" the gear around. Refer to Fig. 2-19.

A rather simple equation gives the theoretical machine advantage (TMA) of a screw. That is the force produced by it on another body or surface. In the case of the worm and worm gear the force is produced on the worm output gear. The equation is stated as

TMA = 6.28 L/p

    L = radius of screw (worm)

    p = pitch, or distance, between turns of screw.

Another expression relating to the screw is that the torque (force × distance) applied to the screw to move a weight W is equal to

$$F \times D = W \, r \, \tan a$$

F = force
D = distance (lever arm)
W = force in lbs the screw produces
 r = screw radius
 a = pitch angle

Note that because the tangent is used, the smaller the angle the larger the value of the left hand side of the equation. When the angle is 90 degrees, the tangent is zero. We might write this equation as

$$W = \frac{F \times D}{r \, \tan a} \quad \text{(the input torque)}$$

Thus, in a simplified sense we may expect the *torque* of the worm gear arrangement to be equal to the radius of the outer gear times the force produced on it by the worm, or screw, section. That would be value W in the above expression. We didn't say it would be easy. Worm gearing is fine though hard to calculate. But you can determine the speed ratios quite easily as you saw in Fig. 2-19.

ACCURACY OF CALCULATIONS

We must point out before we go further that we have taken some liberties (Kryton is now grinning at me—the monster) with the equations used in obtaining our ballpark figures for torque and motor size. In most cases we have tried to simplify these equations so that you might more easily use them, yet still obtain reasonable values as to the size and power required for your own particular robot.

If you are a perfectionist, as many persons (and some robots) are, the question now becomes what to do for better figures and more precise calculations. We can offer three suggestions. First, the local library will have books on gears and gearing and motors in its Mechanical Engineering section. The librarians, we have found, are most helpful and will give you assistance. Also, sometimes a good

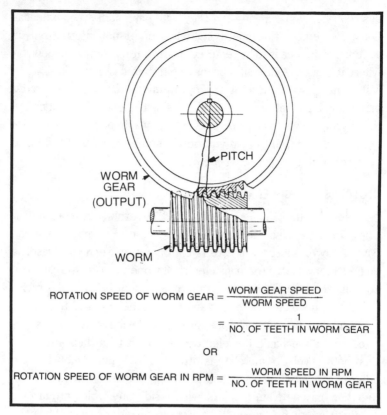

ROTATION SPEED OF WORM GEAR = $\dfrac{\text{WORM GEAR SPEED}}{\text{WORM SPEED}}$

$= \dfrac{1}{\text{NO. OF TEETH IN WORM GEAR}}$

OR

ROTATION SPEED OF WORM GEAR IN RPM = $\dfrac{\text{WORM SPEED IN RPM}}{\text{NO. OF TEETH IN WORM GEAR}}$

Fig. 2-19. A worm and worm gear. If the worm gear is to rotate at one rpm, the worm must turn as many revolutions in one minute as there are teeth on the worm gear.

physics book will be helpful, as they cover mechanics and mechanical advantage devices.

The second method of further exploration is to contact the teacher of the physics department in a high school or the Mechanical Engineering professors at the University. They too will probably be glad to assist you to determine the precise and correct values. They are all *wonderful* people.

Finally, you might write to the manufacturers of servo or automation equipment, or manufacturers of electric motors and ask them to assist you in determining the size motors and gearing you might need. Give them the specifications of your robot—size, weight, speed of motion, desired battery voltage, and how continuous a service you expect—and they will make some recommendations. Remember that specifying whether or not intermittent or continuous service is to be required from your motor may depend on where it is used. A drive motor may operate most all the time, while a steering motor or hand or arm operating motor may be operated only a small part of the time, and for short intervals. We offer the three suggestions to help you prevent the waste of money in obtaining motors and also to help you obtain something which will make your robot all that you want it to be.

CLUTCHES AND GOVERNORS

Now let us examine clutches and governors in more detail. These devices are useful in some applications. For example, if you have a power source which runs continuously, such as a gasoline engine, then a clutch (perhaps electrically operated, as the clutch on your air conditioner on your automobile) might be very desirable. Using it, you won't have to stop and restart the motor all the time.

Clutches are normally pressure disc devices which are separated until a mechanical or electrical force thrusts them together, then both revolve. One plate is attached to a motor element, the other to the output gears or shaft. The pressure which brings them together may be from a coil of wire inside one which produces a strong magnetic field when the coil is energized, or a clutch may be caused to engage by means of a centrifugal force as a motor is speeded up, or it may just be due to a mechanical lever which is caused to move by some kind of force. There must be a high friction between the plates so they won't slip when engaged.

The governor is a device which, in its simplest sense, tries to keep the output shaft or wheel at a constant speed by applying a braking pressure to the shaft. The amount of this pressure is directly proportional to the speed of the shafts rotation. Examine Fig. 2-20, which shows a sketch of one type of clutch and one type centrifugal force governor.

The idea contained in the motion of the governor might be used for other purposes. It might cause some other action to take place when the speed of a wheel or shaft increases to some particular value. For example, it might be used to extend or retract fingers of a robot, since its action is linear. It *could* operate against a spring which would always keep the controlled device at neutral when the motor speed was very slow or off. But notice that the use of this type governor might not be a good idea for motor control because it does consume power by the application of friction, or drag, to the output or motor shaft.

A VARIABLE SPEED ARRANGEMENT FOR SPECIAL APPLICATIONS

It is possible to use two wheels arranged as in Fig. 2-21 to obtain some control over the speed, torque, and direction of rotation of an output shaft. If wheel A is the drive wheel, wheel B will spin at a speed determined by where it is on the radius of wheel A. At the outer edge the speed of the two wheels will be equal, *if* the wheels have the same diameter. Then as wheel B is moved in toward the center of wheel A the speed of wheel B *decreases* and you have the

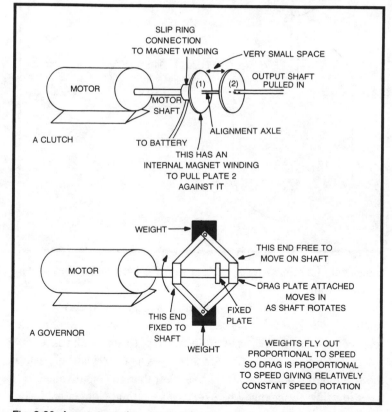

Fig. 2-20. A representative clutch and governor.

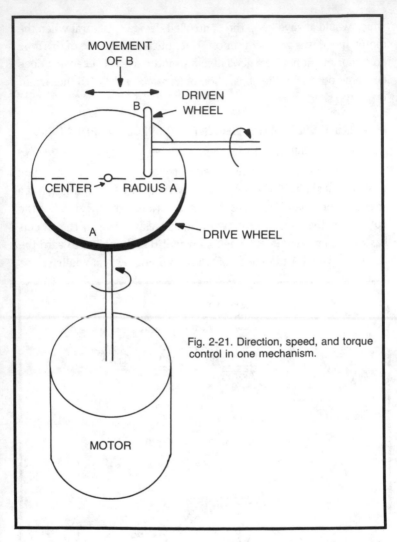

MOVEMENT
OF B

B

DRIVEN
WHEEL

CENTER → RADIUS A

A

DRIVE WHEEL

Fig. 2-21. Direction, speed, and torque control in one mechanism.

MOTOR

equivalent of a step-down speed ratio. This also gives an *increase* in torque output.

If you make wheel B one-half the diameter of wheel A you then can have an increase *or* a decrease of the speed of rotation of B with respect to A, and an increase or decrease of torque with respect to wheel A. Remember that torque increases as the "gearing" makes the speed of the output wheel B go slower than the input wheel A.

How this device might be used is interesting. Notice that if you cause wheel B to move to the opposite half—past center—of wheel

A then wheel B will rotate in the opposite direction to what it was going, but you will still have the same control of speed and torque as before by moving wheel B out or in along this left radius. If this arrangement were used as a drive system, for example, you could have forward and backward motion of the robot, with the speed and power controlled by moving the equivalent of wheel B—or moving A and its motor. At the slower speeds it would have more power to, say, climb a ramp, and at faster speeds, less torque is needed so this would be useful on level ground or floors.

Of course this would require a mechanical arrangement to position the drive wheel on the driven wheel. It also would require a high-friction contact between the wheels. Perhaps there is some kind of gearing—and we won't speculate as to what kind—which might be used to give this effect. We don't know of any as of this writing, but you might think of something and become famous. The use for this mechanism might not be just for drive; it may have a much more important application in some other part of the robot, so keep it in mind.

The arrangement, however, is not as easy to control, from a construction standpoint, as is an electric motor, which can easily be reversed and whose speed is easily controlled by simple electrical elements such as variable resistors in the control line. But never knowing what someone might want to build that might require some special adaptation of speed and torque for his particular use, the arrangement is fascinating to think about.

IMPORTANCE OF THE LOCATION OF THE CENTER OF GRAVITY IN A ROBOT

We mentioned earlier that the pyramid or cone-shaped frame was important because it permits a lower center of gravity (CG). This is important because the lower the CG is, the less likely the robot will fall down, and we don't want that at any time.

If you examine Fig. 2-22 you can see how the CG drops from a higher position to a lower one as the body shape goes from square or cylindrical or round to an inverted cone (or pyramid) shape. In general the CG coincides with the center of mass, although this may not always be the case. Using bisecting lines from the corners, to coincide with the center of area and assuming a constant density, we

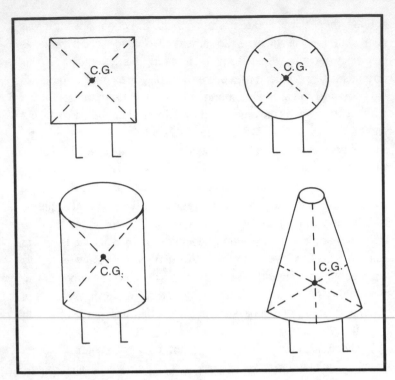

Fig. 2-22. Geometric shapes and the center of gravity.

can have a good indicator of the CG location, which is at the point of intersection of the lines. You can see that a "squatty" shaped cone would have great stability and not turn over easily.

Of course we can overcome a high CG by shifting the mass of the insides. If we make the top part of any body light in weight and the bottom heavy (the opposite of what we get with long legs and a body on top) then our method of intersecting lines for locating the CG does not hold true. If the top is very light then you might ignore it and just find the CG of the heavy mass section by finding its center. You won't be far off this way. Now let's see how the CG fits into some simple ballpark stability calculations.

Begin by examining Fig. 2-23(a). We can assume that the weight of the body is centered at the CG and force F_1 (gravity) holding it on the floor acts straight down through this point. If we have wheels at the base as shown here to illustrate the unbalanced force situation—and *not* to set forth an example as to where the wheels should be located—then we find the stabilizing force to be the product of force F_1 times lever arm l_2.

If we put a weight on the edge of the square such as that shown as F_2, this will act through distance l_1 to turn the body around the axis at the center. Thus you can see that if force F_2 acting through lever arm l_1 has a product greater than the stabilizing force ($F_1 \times l_2$) the body will tip over and fall, in this case, forward.

Now, for the case where force F_2 is not applied, we examine Fig. 2-23 (b) and we see that if the body were balanced on one wheel (or two wheels in parallel) and tilted just a small amount, the turning force acting around the wheel (or foot) contact with the floor would become the product of the weight and lever arm l_2, and the body would fall. Notice that the farther it tilts in falling, the more force is exerted to make it fall as the lever arm gets longer, and thus the torque, or turning, force becomes greater.

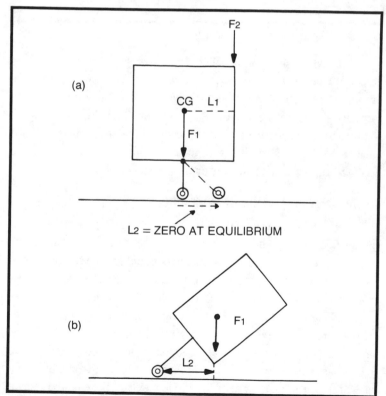

Fig. 2-23. Lever arms and moments.

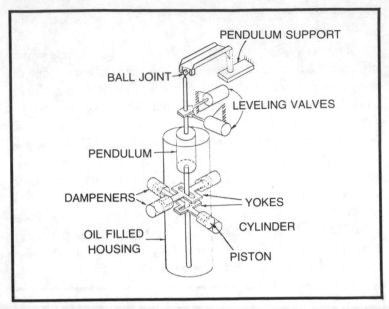

Fig. 2-24. A pendulum leveling device.

This latter situation is called a completely unstable case. It is important because when we try to make a robot walk we think of having two feet in contact with the floor on a line, or parallel, as perhaps similar to the wheel shown in Fig. 2-23. You will always want to have the robot body sturdy and stable enough so that minor pushes or attempts to tilt it won't cause an upset. You want a good footing support and a low center of gravity—always.

DYNAMIC STABILIZATION OF ROBOT BODY

This idea concerns the use of something like gyroscopes or pendulums, such as shown in Fig. 2-24, to keep the robot from falling over. These would require some kind of counterforce to oppose the tipping force. With a gyroscope it could be the *precession* force; we won't expand on that here. With the leveling pendulum, the mass of the pendulum might be caused to swing to oppose the tipping force, but damping (to prevent wild swinging of the pendulum) is necessary, and it is hard and critical to adjust. The device shown in the figure was used to level a four-wheeled device. The leveling valves applied power to pistons to cause the low side of the cart to raise on slopes.

MORE ABOUT "WALKING" TOYS

These "walking" toys are fascinating and since there will be more and more of them they warrant inspection. Some of the ones examined and found to be of applicable interest were already shown in Fig. 2-2 and 2-9.

Refer again to Fig. 2-9. This not only shows the gearing used to power the output wheels, but also how the simulated foot and leg are attached to the output wheel so that it moves in a walking-like motion, although the foot never comes in contact with the floor. The forward motion is accomplished entirely through the use of the pair of large wheels which, with a third stabilizing wheel in the gear, make it possible for this robot toy to move along a straight-line path. The linkage to make the foot move back and forth and up and down are worthy of notice and might form the basis for some other kind of walking machine if some difficulties are removed.

Why not use this kind of walker? Well, notice that if you let the foot come down and raise the wheel which moves it, it would push all right, but it would have a turning tendency. That is, it would tend to turn the toy in a sideways manner. When the opposite foot came down it would tend to turn the toy in the opposite direction. Thus you could have a waddling effect which might not produce much forward motion at all. That brings up a fundamental question as to whether this mobility would by any good at all. *You* might find an answer, providing the turning motion can be eliminated.

Referring to the other toy which does "walk" Fig. 2-2, we again recall that gears are used to power the crank type output. This will raise and move forward and backward a linkage which is likened to a leg. Each is very large, considering the toy size. They are so designed that *each* will give balance alone when supporting the toy body. Each foot is also slotted so that one can rotate within the other. These feet keep the body from falling too easily although it will fall over quite readily at times.

Keeping a balanced force on a wide-base support such as shown in Fig. 2-25 helps to prevent toppling. Here we see that there is a restoring, or balancing, force which is equal to the product of the lever arms ($L_1 \times W$, and $L_2 \times W$) which will keep the body from rotating about point X. Thus it wouldn't fall if the lever arms are long

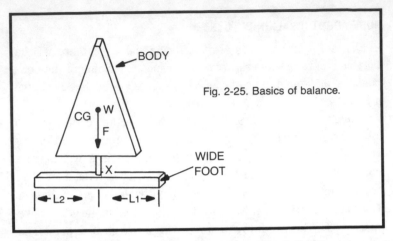

Fig. 2-25. Basics of balance.

enough in each direction. To keep the body from falling into or out of the paper, the foot lever arm must be made *wide* enough so that the lever arms times the weight also gives the large balance force necessary to prevent tipping.

As the feet rotate up and over and down and up and over and down there *is* a slight unbalance created in the body itself. And at times with this toy, depending upon the amount of unbalance, the tipping force may exceed the balancing force. Thus, the body may tip over. However, for normal and level use with slow, constant motion without jerking, the body will not tip over, and does walk forward.

This is an interesting toy because, again, it brings up the possibility of a walking robot. He would have very large feet, that is certain, and he might be short and squatty but he might also get around pretty well too. We showed in Fig. 2-1 a close-up of the leg and foot which is attached to the small crank. Notice that the feet in the side view plane are so raised that one, when moved forward, rises above the other completely so that it doesn't act on the body until it comes down and raises the opposite foot off the surface. Each foot provides an independent balance.

There is probably a limit to the size of the feet you could use on your robot. If you get them too large (long) then they may not rotate up and over cleanly, but would tend to drag, and that could be disastrous. But then, you might, in your own dreaming and experimentation, think of a way to hinge the foot and drive it through an additional linkage so that it could bend as our feet do, at the ankle, and thus avoid any dragging that would occur with the rigid-foot construction.

GEAR BACKLASH

We must not forget to mention, before we proceed further, that you should always consider backlash when you work with gears. This simply means that some gears may not mesh tightly, so there is some play between the teeth. The result of this is that the gears may not properly position an arm. Or more importantly, a feedback instrument, or as it's sometimes called, a transducer, may not be positioned properly. This may result in hunting or some other instability in the operation of the robot.

We want tight gearing, without play or backlash, so that when the parts are in operation they move smoothly and evenly without any sloppiness due to gears or other mechanical parts being loose with respect to one another. Notice that this includes linkages, which are hard to put together so that they operate smoothly and easily without much friction, yet have very little or no sloppiness.

In some precision mechanical devices gear pairs may be used, one fastened to the other with a spring arrangement, both gears being identical and on the same shaft. The spring arrangement always keeps one gear tight against the drive gear and thus eliminates sloppiness and backlash. The spring is usually internal to one gear and is simply fastened from one gear to the other and operated under spring pressure against the drive gear.

We mentioned play in pulleys and belt drives, and we want to emphasize a point here when using them. Be sure to check belts frequently for hardness, drying out, cracking, and proper tension. Check the pulleys for wear and tightness on their shafts. You should always design your system so you can make these checks without undue effort. Replace the worn or bad parts as soon as they are noticed.

STEPPING MOTORS

These are somewhat special but deserve mention as a means of propelling a robot or making the various parts of a robot move. These motors operate by electrical pulses and "step" forward or back a small, but specific, amount with each pulse applied. This means that if you apply the pulse signals continuously, and each pulse steps the armature 3 degrees, then you will find that it seems that the armature is simply undergoing a continuous rotation. But really it is moving a very precise amount with each pulse. Because of this precision rotation, a very fine degree of control can be obtained over anything which is driven by one of these motors. These motors can be synchronized to have exactly the same speed of rotation. As stated, you can reverse the direction of rotation by changing the polarity of the pulses applied.

Now consider the application to a robot drive system. You can step one motor forward and step a second motor backward to give a very quick and tight turning capability. Also you can adjust the pulses to each of, say, two drive motors to get steering as well as propulsion. If you send a few more pulses per second to one wheel, then the driven device will turn in the direction of the slower wheel. Gradual turns are accomplished by small differences in pulses, quick turns are accomplished by large differences and very tight turns can be accomplished as mentioned previously.

It is very interesting that with this kind of system you have both steering and mobility combined with two drive wheels, just as with the electric wheel chair. The only consideration is that you must apply pulses to the motors instead of a continuous voltage, and the pulses must be of a variable rate and adjustable polarity. You might want to contact some motor manufacturers to find out more about these kinds of propelling and steering devices. Also you might want to use them for arm or finger movements where fine precision is desired. Write to Winfred M. Berg Inc. 499 Ocean Avenue, East Rockaway, N.Y. 11518 for a complete catalog of precision gears of special interest to hobbyists.

CHAPTER 3
SOME PRELIMINARY
CONSIDERATIONS OF SENSORS

If a robot is to be self-activated, it must respond to an internal program. This might be initated by a sensor input. The sensor is some device which converts a physical phenomenon such as light, heat, sound, radio waves, magnetism, or human presence into an electromechanical system inside the robot into a physical or other action. When we say other we mean, for example, the activation of a light, laser, sound, or whatever. Some action will result from a sensor being activated.

Now when we consider the robot's basic system, we must realize that this can have various parts of it activated by a timer. This can be an internal programming device such as a clock with many button switches so arranged that as the hands move round the face, they cause electrical contacts to open and close at the various times and for specific lengths of time, as we have arranged it. The timer thus may cause the robot to move, raise its arms, flash its eyes, speak, and emit a light beam as a function of time once this program has been initiated by some appropriate sensor. A simple timing mechanism does the ordering of tasks, and it orders them by closing switches at various times and opening switches at other times. This kind of programming is basic to robotics. We will learn more about this very fundamental and important device as we proceed further. It will take a multitude of physical forms, as we shall see.

LIGHT SENSORS

But, back to the input device, the sensor. Let us begin to consider what kinds of physical phenomena we might expect our robot to encounter, and then think of what kinds of sensors will be able to detect and produce signals from these pnenomena. Of course we think immediately of sound and light. Sensors for these pnenomena are available and well known. For sound a large or small microphone, and for light a photodiode, phototransistor, or photocell.

The first two are solid-state devices which react to light falling upon their juunctions. A photocell produces electricity when exposed to light. An example of these devices is shown in Fig. 3-1.

One manner in which such sensors might be used is shown in Fig. 3-2. If we set a prism (a), a triangular-shaped piece of glass, in the manner shown at (b) and place a photocell near each side as shown, then when a light source is directly ahead, perpendicular to the prism surface, the light will enter the prism, bounce from one side to the other and go back out in the direction it came. Very little light will fall on the photocells.

Now consider what happens when the light is moved to one side as at (c) or (d). The light then enters at an angle, is refracted in such a manner that it goes through the prism and comes out on one photocell much stronger than it does on the other cell. So we could connect a steering device to these cells such that the one receiving the most light would make the robot turn to face the light source.

If the light were moved to the opposite position then the other photocell would receive the most light, making its signal the strongest. Through the control mechanism, the robot could be caused to turn to face the new light source. In this manner then, we can make our robot seek or follow a light source, be it a light or light rays reflected from a prepared path. More about that in later pages.

SOUND SENSORS

Of course we are all familiar with microphones used with CBs and public address systems. We know that they may be easily obtained in large or small size. In fact, some may be so small that they are hardly visible to the eye. These are the types used in

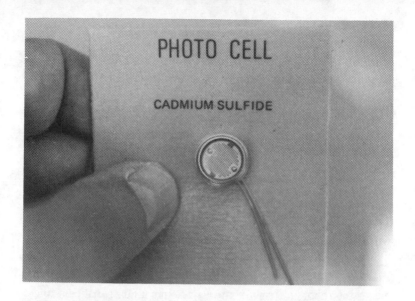

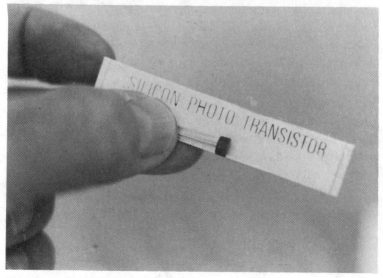

Fig. 3-1. A photocell and a phototransistor.

hearing aid glasses. They blend in so perfectly, and are so small in the frame, that it is difficult to tell they are there. Thus, when desiring an illusion such as the robot hearing when you want it to respond to some words or some noise, you will want the sensor to be as small and hard to detect as possible.

There are many devices now on the market using microphones as sound sensors to initiate specific programs. One such device is a small toy automobile which, once the on switch has been closed, seems to respond to the spoken word. You would make up a sound code such that, for example, "Go" makes it go forward, "Turn left" makes the car turn left, "Now turn right" results in it going to the right, and finally "You must now stop," stops it. This would likely indicate that you have a code consisting of 1, 2, 3, and 4 sound pulses, if each word is spoken with care and distinctly. These sound pulses can be used to activate a steering and control mechanism, such as a four-position escapement in the car, to make it respond to the code. This code system was used with the Westinghouse robot Elektro, which we discussed in chapter 1.

When we use a *sequence* control system, which, incidentally, is easy to construct, it might be arranged this way: The first sound makes the robot go straight, the next sound it hears makes it to go, say, left, the next sound it hears causes it to go right, and the fourth sound it hears makes it go backward. The sequence then repeats in that order. This means that if the car (robot) is going straight and you want it to back up you must say something with just the right number of words, for example:

"Go"—first sound causes forward motion.
"Turn"—second sound causes turn left.
"Turn"—third sound makes it turn right from left.
"Back"—fourth sound makes it reverse.

Then to go from forward to reverse you'd say "Back up now," which consists of three words or sound pulses. If you think about this you'll find that it can be fun, but might present some problem to you remembering how many words to use.

In order to go from left turn to forward will require three words, but to go from right turn to forward requires just two words. You'd have to be pretty alert and remember when to speak one, two, three, or four words to make the robot (car or whatever) do what you want it to. But it could be done, to the amazement of your family and friends. What we really want, however, is a system which does not have a sequence. You tell the robot what to do and it does it, and you don't have to remember any "code" or anything like that. It

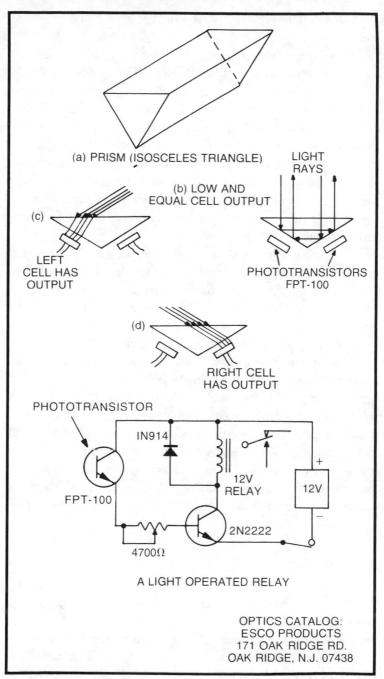

(a) PRISM (ISOSCELES TRIANGLE)

(b) LOW AND EQUAL CELL OUTPUT

LIGHT RAYS

(c) LEFT CELL HAS OUTPUT

PHOTOTRANSISTORS FPT-100

(d) RIGHT CELL HAS OUTPUT

PHOTOTRANSISTOR

IN914

12V RELAY

FPT-100

2N2222

4700Ω

12V

A LIGHT OPERATED RELAY

OPTICS CATALOG:
ESCO PRODUCTS
171 OAK RIDGE RD.
OAK RIDGE, N.J. 07438

Fig. 3-2. A phototransistor with prism.

should respond to your commands no matter in what order they are given.

Sound can also be used in a more subtle sense to determine objects and how far from the robot these objects might be. Like radio waves from a radar, sound can be reflected from surfaces, and certain frequencies of a sound reflect better than others. Higher frequencies are better for this purpose than the lower frequencies. We define the operating spectrum as from zero cycles per second (hertz) to 50,000 cycles per second (hertz). In order to use these higher frequency sounds like a radar uses its radio waves to determine distance to objects, the sound system would have to be designed like a radar, with appropriate directivity, timing, and comparison circuits.

But sound can be used for control, and many of these systems have been developed. One such use is for the guidance of blind persons. Whether it is better to use an actual miniature radar or to use a sound radar in your robot (which you might say is like the Navy's sonar systems in miniature) may be determined by cost, availability, and complexity of the systems involved.

There is one other aspect of sound that we need to be aware of, and that is the speed of sound. Sound waves move at about 700 miles per *hour* compared to the 186,000 miles per *second* that radio waves travel. Thus there may be situations which will prevent the use of sound waves because they travel too slowly. Also, the environment may prevent the use of sound waves because interference signals of the same frequency may be generated by any number of means. This may cause a malfunctioning of the robot system using them for activation. Garage door opening systems use special codes to prevent this problem.

STRAIN GAUGES

Strain gauges are another type of sensor which is important. They are resistance units which are made solely for the purpose of sensing a change in the strain or stress of a mechanical body. Aircraft designers use these, for example, to determine the bending stress on wings and body parts of new aircraft, or on other aircraft which may show fatigue at various places, seemingly without cause.

A simple strain gauge you can make is illustrated in Fig. 3-3. It is simply a resistance which is so arranged that when its base is deformed in any way, a change in resistance occurs. If the base bends in one direction the resistance may increase. Bending in the opposite direction may cause the resistance to decrease. Thus the change in resistance shows the amount of stress, or bending, and the fact that the resistance increases or decreases indicates in which direction the bending takes place.

The strain gauge can be used in a balance bridge circuit such as the Wheatstone bridge in Fig. 3-3(b). Then the bending can result in an actual voltage output from the circuit, which can be positive or negative polarity. This indicates the change and direction of the strain in the element to which the gauge is attached. Sometimes these gauges, which can be the size of a postage stamp, work into a circuit in which they control the frequency of an oscillation—the circuit is a voltage controlled oscillator. This is true in telemetry applications, where the frequency change is transmitted from a remote point by radio, then is sent into data reduction equipment to analyze the strains indicated by the changing tones.

A strain gauge can be used as a sensor in a robot to give a feedback as to the amount of pressure the robot is exerting with its mechanical fingers, or the amount of force it is exerting to lift an object. It might also be used to indicate stress in the robot body or frame if the body is tilted too much or caused to assume some position which is not normally used in its operation, for example, a tilt from the vertical under a load. The voltage output from the strain gauge circuit might then be used to cause a stabilizing system to go into operation to keep the robot erect.

If the strain gauge is used on the fingers, which might be specially constructed to bend a small amount under pressure, then this signal from it could be sent to a controller which controls the motor causing the fingers to apply the pressure. Thus the fingers will not apply any more pressure than required, even being so sensitive that the hand and fingers can grasp and pick up an egg without breaking it.

Let us imagine some of the mechanics required if the robot hand is to lift an object which it must also grasp and hold firmly. We assume that some program has initiated the action and caused the hand to be in the proper position and the fingers closed on the object. Now, if the arm lifts and, say, there is no increased weight on the arm, meaning that the robot has the object in its hand, then the object probably has slipped from the grasp and is not being held. The strain gauge can determine this as an added weight or strain on the arm at some sensitive point. So now the arm again positions the hand and again the object is grasped, but with a little more pressure, and the strain gauge on the fingers will indicate the fact that an increased pressure is being applied. Again the arm will raise the hand, and if the object has been grasped firmly enough—but not too firmly—then the lifting action will have been accomplished. Of course the hand must have some kind of flexible and nonslippery material such as a rough plastic or rubber so that it can grasp smooth objects and hold them. In any event the try-and-lift-and-grasp-again procedure will continue until the robot actually picks up the object with absolutely no more force than is required to keep it from falling or slipping out of its hand. After all, isn't this basically the way we humans do it? Think about it.

THE PIEZOELECTRIC CRYSTAL

This device is another possible means of providing information as to pressure or bending. These tiny flat crystals placed in the fingers of a robot and connected to an oscillator circuit where they control the frequency will cause a change in that frequency when pressure is applied to them. Of course it may be true that this means would not be sensitive enough, or that it might be too expensive compared to other systems of force measurement, or that it might be too complex to use. But we mention it because it is an actual means of sensing pressures, however slight they may be. The idea has been used as we shall learn in some later pages.

MOTOR CURRENT

We might not think of motor current as being a sensing device or being sensitive to forces, but it is. Perhaps it isn't directly sensitive to forces, but the increase or decrease in drive-motor current due to resistance (or forces) on the output mechanism (gears) moved by the motor *could* be used to sense the force or resistance encountered. Some quick thinkers will immediately look

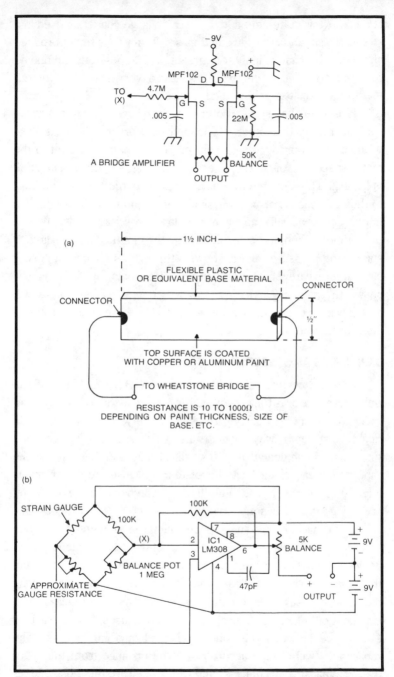

Fig. 3-3. A form of resistance strain gauge appears at (a). This sensor could be used in the circuit shown at (b).

askance at this statement and wonder if a large gear train used in a step-down situation would reflect so small a force as to not produce an appreciable change in motor current. This is possible, but maybe the gear train won't be that large, or a very sensitive current measuring circuit could be used.

It is true that some motors will produce more measurable current changes than others. Also, some motors may have a special winding just to produce a current or voltage output related to the motor speed, and of course, speed may change slightly as a function of loading. This idea, anyway, is a possible source of information as to the resistance (grasping pressure) of an output hand driven by a motor and especially applicable to large-load situations. With a proper motor then, it is possible to have the motor do double duty. It applies the pressure, and it shows—through its change in speed, current, or whatever—how much pressure is being applied. It might be worthy of more investigation and study. Various motors will certainly have to be looked at, as some will be more indicative of this effect than others.

CAPACITORS AS SENSORS

One very common electrical component which might be used as a pressure sensor is the capacitor. It can control the frequency of an oscillator simply by changing the spacing between its plates. This might be a dandy way to generate a pressure feedback signal. Fingers might be designed with small capacitors in them so that the capacitance can change as the plates get closer together under increasing pressure. (You might have to watch out, however, if you are trying to pick up a nonbending body. You don't want to squash it.)

The capacitor would have to be built into *each* finger so it varies this way. The oscillator which each one controls the frequency of can be designed to have a linear range over the changes, thus providing the information needed. The capacitor hand would have to be strong but capable of slight deformation under pressure, and limited so *it* isn't bent out of shape. Capacitors have other uses. There will be a capacitance between the robot body and anything near it. The capacitance will change according to the distance from the robot body. Humans, metal objects, and wet ground are the most influential objects which can cause the robot body capacitance to change.

98

Now, if the robot body is connected as a small antenna to a small oscillator, this capacitance change between the body as an antenna and nearby objects can cause an increase or decrease in current of the internal oscillator. This, in turn, can cause either solid state or mechanical relays to operate other subsystems or programs. For example, one system might be just a backing away movement of the robot. Another might activate a voice recording and speech circuit; another might have the robot wave its arms and flash its eyes to frighten away persons.

But the basic activation simply would be the presence of anything that will cause the tiny current in the oscillator to change. The antenna itself might be a little green man antenna sticking out of its head. Or the body itself as we mentioned. You have seen lamps in stores which turn on and off when touched. This system might also be used, but touching is required and you might not want anyone to actually make physical contact with your robot.

RADAR SYSTEMS

Many of us know about radar. If you don't, we recommend (of course) our book TAB No. 575, *Modern Radar: Theory, Operation, and Maintenance.* These electronic transmitter units send out pulses, or signals, which are reflected back to their source. The time it takes for the signal to get there and back is measured to determine distance to the object or reflection. Since the signals must be reflected, they must be strong signals. They often must be reflected from some object which permits nearly complete passage of the signal through it. By measuring the elapsed time between transmission and reception in millionths of a second, and by knowing the speed with which radio and radar signals travel at, it is possible to calculate (automatically with a built-in circuit) how far away the reflecting object is. Radars are also designed to determine how fast an object is moving (police radars) or they can determine is something is moving without regard to what it is as in the case of security radar systems.

This means that radar is another possible sensing device applicable to our robots. It used to be true that the size of a radar system made it impractical to use in a robot, but that is not true any more. Visit your local Police department and look at the radars which are

BELMONT COLLEGE LIBRARY

used to monitor motorists. These units are very small and light-weight, considering what they do.

A radar type sensor which uses any of the three ranges of radiation—SOUND which has the lowest frequency, RADIO OR RADAR in the midrange, and light, INFRARED, and X-rays in the high range—are adaptable as a sensor to determine the position of things with respect to the robot's position. There are two requirements which must be met. There must be a transmitting antenna, or the equivalent of this, and a receiving antenna or the equivalent. Sometimes, with special circuitry, one antenna can serve both functions. Even in units operating with very low frequencies it may be possible to use the same output device for transmitting and receiving this low frequency energy. For example you might think of an intercom which uses the same speaker for receiving and sending sounds.

As the sound frequencies go higher, necessary to get our desired directivity and to get the sound above the hearing range of people, it may be necessary to have a special transducer to send out the signals and another to receive them. Some TV sets use this higher-than-audible sound range to control channels, off-on and volume, and so forth. A hand-held transmitter sends out a sound signal, which we cannot hear. There is a small receptor unit on the front of the TV to receive this signal and send it to the circuitry. Usually the transmitter must be pointed nearly at the receptor, or the signal won't be strong enough to activate the required operation. Figure 3-4 shows one example of such units. Note the small size of the transmitting unit.

THE LASER

The use of a very narrow, intense light beam to obtain direction and distance from a robot is not impractical since the laser has been invented. But beware of its use as it might blind a person who gets the laser light in the eyes. The laser sends out such a narrow beam of light that it can pinpoint an object. The light reflections from the small area illuminated can establish positive distances. The circuitry using light is like that of the radar previously described.

If you used a flashlight to send out a beam of light, the beam spreads through too wide an angle. Thus a photocell trying to

Fig. 3-4. This TV controller uses high-frequency sound to transmit to the rectangular "antenna" on the set.

determine direction from the reflection would likely be confused. Position determination using a wide beam of light is just not very accurate. A camera electronic flash, or strobe, in the more expensive models has a thyristor which is adjusted to look at the light being reflected when the flash goes off. It is so wired into the flash circuit that it controls the amount of light being emitted, thus keeping the light at the correct level for proper picture taking with a given lens opening. But notice that this kind of sensor acts on the average of all light being reflected, and while it acts very fast, as it has to in order to control the amount of light being produced, it cannot tell how far it is from the reflecting surface. The thyristor merely controls the level of the light. But think of this idea and you may think of some means to use this system in your robot. Visit a camera store and examine their automatic strobe systems.

Let's consider using strobe sensor with, say, a laser. Consider the thyristor being wired into a circuit which will produce a pulse output signal when it gets the reflection, instead of controlling the amount of light emitted. Now you have a "light radar" which *can* determine distance and objects, just as a microwave radar does. Of course the range won't be as great, but for a robot, range may not be

101

a problem. If you can cover 50 feet that may be sufficient for most purposes. If the robot is to act as a sentry at some installation, then a longer range will be useful, perhaps a half mile or so. Finally, in some military applications the laser is used along with light-sensitive sensors to control the direction of a Smart Bomb. It is a feasible system for use in a robot.

MAGNETICS

There are sensors which can be used to indicate a magnetic field, or the presence of objects within a magnetic field that change the field pattern. You have probably not seen these directly, but they are used in airports at the boarding ramps, built into arches through which you must pass to reach your plane. If you have any large metal object that disturbs the field, such as a big bunch of keys, an alarm will sound. The arch has a magnetic field through it. When you pass through this field to go on the airplane you are examined. Any change in that field, due to a metallic presence (humans don't change the field), is detected by the sensors. They do this, perhaps, by detecting the change in current to some kind of an oscillator. This change then, in turn, causes a relay to operate which causes the alarm to sound.

A robot might be made to avoid metallic strips which could be placed in spots along its route, thus telling it to stay away from places we didn't want it to go. The robot could generate a magnetic field which would, in the manner described previously, activate relays, causing it to move opposite to the direction that the field disturbance comes from. He moves until there is no longer any disturbance.

Another way in which the magnetic field might be used is to use coils as sensors (they are also transmitters of magnetic fields if iron or ferrite cored) which can pick up the tiny field, or signal, from wires carrying alternating current at from 1,000 to 25,000 cycles per second. If some robot system can sense this signal the robot can be made to follow the wires. They can be buried in the yard or placed under carpets in the home. This type of system is not too complex. Look at Fig. 3-5.

When two magnetic sensors are used such as shown, then as long as the signal pickup in both sensors is equal the robot will move along the cables in a straight line. If the signal pickup is stronger in

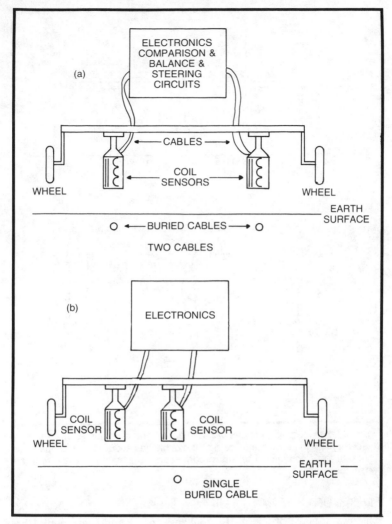

Fig. 3-5. Magnetic sensing of two cables (a), and magnetic sensing of a single cable (b).

one sensor than in the other, the robot should move away from the stronger signal and toward the weaker signal until balance is again restored. A simple comparison amplifier of the integrated-circuit variety can make the comparison, and there are many available. This amplifier would then operate a polarity sensitive circuit and relays which would cause the proper directional control to be made in the robot's movement. A system using three core sensors is shown in

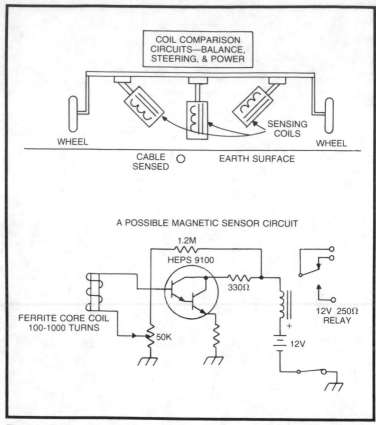

Fig. 3-6. Three magnetic sensors and one cable. Three wound ferrite cores can detect a 1-kHz signal in a buried cable if the cable has about 100 milliamperes of current at this frequency. The return cable can be spaced far away from the sense cable.

Fig. 3-6. Of course you might use two sensors to sense a single cable carrying a signal of about 100 milliampere strength at about 1000 kHz, as in 3-5(b).

THE LIGHT SYSTEM AGAIN

This brings to mind that a light path might be used the same way as the magnetics path is used. If you paint a white stripe perhaps an inch or so wide on a floor, and then beam a light down on it from a close distance, properly adjusted and focused, you can get a good reflection back from this strip. If you then place two photocells to produce currents proportional to the light intensity falling on them, one on each side of the light, but shielded from the light by a tube so arranged that the light sensor looks only at the white strip. Then, if these cells are properly placed you will have another means of providing your robot with a path control system. The operation will be similar to the magnetic field path system just described. The light sensors feeding the same type comparison amplifier etc.

Now consider some differences in these two systems, the magnetic and the light path. In the first case using the magnetic system the path won't be visible as it can be concealed. In the second case the sensor must "see" the white line and thus you and others may see it also.

Finally there is a path system which might use radioactive particles which have been placed along a line where the robot is to travel. Radioactive sensors are available and are used in some applications. The system would be physically like the other two just described, but here there would be no visible path, no buried wires, and so the robot just seems to be going on its own. How about that!

RADIO BEAM PATHS AND LIGHT BEAM PATHS

You might ask about creating a radio beam path for the robot to follow, or you might think about having the robot follow a distance light source which might move or be fixed. Tracking a light is not difficult to accomplish, but to have the robot follow a radio or radar beam might be more difficult. The size of the beam produced by light, radio, and radar waves is inversely proportional to the size of the antenna (or focusing system) beaming it. That means that to create a tiny, slim, cone shaped beam requires a very large antenna, or much optics. The laser mentioned earlier, of course, is an exception to this rule for light, although its beam does spread a very small amount over large distances.

But a robot *could* be designed to follow a light source if the beam were narrow enough so that the robot didn't wander back and forth across the path, shown in Fig. 3-7, as it seeks or tracks the light.

The advantage of using this system is flexibility in path position. You simply change the light source from here to there and the robot will go there. But the complexity and possible required size of the antenna in the case of radio or radar waves would be too great unless a very high microwave frequency were used.

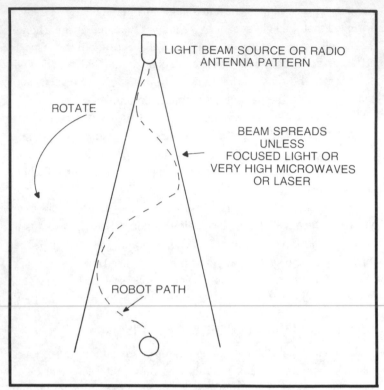

LIGHT BEAM SOURCE OR RADIO
ANTENNA PATTERN

ROTATE

BEAM SPREADS
UNLESS
FOCUSED LIGHT OR
VERY HIGH MICROWAVES
OR LASER

ROBOT PATH

Fig. 3-7. A robot can be made to home on a light beam or radio signal.

Consider a photocell, Radio Shack type 276-116; with a light source provided it can be used across a doorway to announce persons entering or leaving the room. Now consider buying *two* such receiver units (already built). Now with the light source you have a beginning system for light tracking. Place the two receiving units just far enough apart on the robot so that they are not activated when the light beam is centered between them, Fig. 3-8. You have the essential system to guide the robot to the light source or away from it. The photocell receivers are the robot's eyes. They have to be placed some distance apart on the robots head (or elsewhere). The output of each receiver which has a built-in relay can then be used to apply the proper polarity voltage to the steering motor to cause the robot to try to keep itself centered in this light beam. Of course in bright daylight and over a great distance the problem of which light is which becomes a problem.

Of course we might consider this same system with the light source placed on the robot. A high intensity pulsed light such as used for marine emergency conditions, or an aircraft type high intensity pulsed light might be considered. The robot then might be able to look at and direct itself toward an object which reflects the greatest amount of light. A camera strobe is another source of this kind of light, but you'd have to arrange a circuit which would pulse it at regular intervals. Such a circuit is a timer connected to a mechanical relay. Such timer circuits are available at radio parts stores, and camera flashes are available at camera stores in new or used types.

This kind of pulsed light system, however, has a very serious disadvantage. It might blind anyone who happens to be in front of the robot. No one, not even the gremlins would want that. It's not as bad

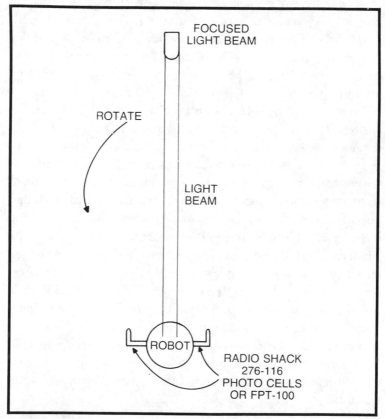

Fig. 3-8. If this robot strayed from its path tracking the light beam, the appropriate light sensor would guide it back on course.

as a laser but it's pretty bad because of the high intensity of some types of flashes. This type light system, using pulsed light, might be better used with a white stripe line. This will give good reflection, provide good path control, and in all probability, will operate in daylight. It won't be seen by observers if you have a low skirt on your robot. The light system operates under that.

ELECTRONIC SENSORS FOR RADIOACTIVE SOURCES

We must consider in sensing devices the possibility of using something like the sensor of a geiger counter, which is used to detect radio activity. This is another off-the-shelf item. It has been used extensively in exploration for uranium. The unit is not too large, self-powered, and designed to be hand carried. It is sensitive to small sources of radio activity, such as from the wrist watch on your arm, if you have one which gleams of its own accord at night.

Now a geiger counter, or rather *two* of them, or two of the sensors which are used with the unit, each sending a signal to a comparison amplifier, can be used for path determination. You might want to have a path which is invisible to the eye and which has no bulky parts such as wires. A radioactive chemical might be the answer. If this chemical is impregnated into a rug or carpet in a thin line, for example, it wouldn't have enough radiation to harm a human and yet it will provide a well defined path which can be detected by the geiger counter type of sensor. As you know, the counter shows, on its meter, the intensity of the radiation. Thus we can feed this voltage or current output into the comparison amplifier instead of the meter, where it can be amplified and used to control some relays which give the robot steering. One unit must give a positive voltage output and the second a negative voltage output (which can be arranged easily). Then the steering control direction is specified. A steering system which turns the robot in one direction with a plus signal and in the opposite direction with the minus signal, and in an amount in either direction according to the *level* of the signals makes a perfect steering unit. A small voltage then would mean a small correction and a large voltage would mean a large correction, as an example.

There is, in the Pentagon, so we are told, a robot which doesn't look like a robot at all. It is a mail carrying and pickup cart, but it operates autonomously (by itself) and it follows a chemical path in the carpet as it steers itself down the aisles around a prepared track. It is programmed internally to stop at various points for a specific time so that mail, or whatever, can be taken off or put on the cart. From what we have learned, it is a very successfully operated device. It has great stability, being a four wheeled cart. Its battery lasts about 48 hours between charges. It *could* use the same kind of comparison of two signals from two sensors as we have described earlier. It moves slowly enough so that corrections to its path are almost impossible to detect. As it moves it just seems to move alone easily and positively and holds its middle-of-the-aisle path with no problem at all.

DISTANCE SENSORS

It seems appropriate here to indicate how such a unit can be made to stop at precise distances along its path and then start again after a preprogrammed time interval. Of course one might arrange a special sensor and an isolated bit of radioactive chemical which would initiate a delay circuit to stop and then restart the cart. Or it might use a clock. Yes. The latter is a possibility. With a given speed and a constant speed, it will travel a given distance every second. So it is not impossible to use a clock to let it run forward for a given time, stop for a given time, and then restart and run forward again, etc. One problem with this concept, however, is that starting and stopping—especially starting—may not always be at exact time periods to regain moving speed. It doesn't reach its constant speed for several seconds and so, perhaps, the timing may get out of step with the positions. It may, in time, creep past the place it is to stop, or stop too soon, to the utter frustration of the person waiting with a heavy package to place on it. A better way (better mouse trap?) perhaps, would be to provide one wheel with a revolution counter and sensor which will convert every turn of the wheel into one pulse. Now, knowing that the linear distance traveled by the wheel in one revolution is $2\pi r$, where by r is the wheel radius and π is 3.1416, you can say that one pulse is exactly a given number of inches of travel. Thus you can specify the stops in wheel revolutions. Or, if this isn't precise enough then cause *four pulses* per revolution and divide the wheel distance traveled by four.

However you do it, with a simple counting circuit, the machine will go a given *distance* and stop. The speed becomes *un*important. Of course you might equip the machine with stepping drive motors; and since the effect is the same, that is, the cart moves forward precisely a given distance with each pulse to the motor, you know exactly how far it will go with a given number of pulses. The operation, however, with a wheel pulse-generating circuit and the pulsed motor system will require a timer to hold the machine in any given stop for the required period of time. But this can be done by a constant program operated by pulses which it counts between stops. This effect and system might be thought of as you think how you are going to vacuum the den and living room.

Distances around turns would have to be carefully calculated *at the wheel which is doing the measuring*, if you use the distance measuring system. There is some question as to whether an exact distance can be determined rounding a turn, which can be consistently followed each time the robot goes along its path. But even then, if it is following a path, invisible or visible, it should be close to the line and the circle segments of turns which you have laid out and preplanned for it to follow.

TELEVISION SENSORS

The classic concept of a mechanical robot pictures it with television eyes, microphone ears, and a speaker mouth, to mention a few similiarities to a homo sapien. Let's think about that television system as an "eye" sensor, and we find that there may be some difficulties because of the way a television camera normally works. It is large and complex although there are some portable units, of course. Let's imagine the simplest kind of television camera.

This unit scans the field of vision and then charges and/or discharges a grid of light sensitive cells according to the light pattern it sees. This was briefly mentioned in chapter 1 and you might want to reread that section and examine the concept again. But realize that the picture may not be perfect with this means—that is a matrix of light emitting diodes which are lighted in exact agreement with the intensity of the light falling on the camera grid system. But what is important is that using a grid system such as this you can devise a system where a certain amount of light causes a voltage to appear—and anything less than that amount of light will not cause any voltage. When you can have a output condition of either voltage or no voltage you have an input signal for most computers—a binary information system. You have either one or the other of only two possible conditions, a voltage or no voltage.

This concept then leads to the possibility of pattern recognition in, perhaps in a home environment. If the robot is moved along a prescribed path it can store in its computer memory the light and shadows of what it sees along this path as binary information; then using this record the robot's integrated circuitry can compare at every instant what it sees with what it should see. If the pictures don't match, it will energize another program inside itself which will cause it to move slowly to the left and then to the right until it does get a match. When it gets a match, the steering program holds this and the robot continues forward at the prescribed speed. Notice that we have implied that the robot slows down when it is trying to find out where it is. This might be important as we don't want it to run wild trying to match shadows.

The television concept will require a computer to resolve the information from its camera sensor, but with all the types of computers on the market—including some that you yourself can build from kits—that should not really present a problem. We will discuss the programming and operation of some types of computers in the later chapter called The Robot "Brain." There we will give enough information on computers so that you can understand what is needed and generally how they work and how to go about experimenting with that basic and important device.

A VERTICAL SENSOR

We humans have in our inner ear a tube which is filled with a liquid and in which many tiny hairs, or sensors, are present. These sensors determine the position and motion of that liquid. Our brain then interprets this data into whether we are moving, and *how* we are moving, as well as to whether we are upright or tilted or bent over or whatever. It might be that our robot would need to be able to determine this kind of information. We don't think a tube like we use in our ear would be practical because the liquid moves in response to motion, and if it is too thin a liquid it could cause the robot to jiggle, or hunt, its vertical. If the liquid is too thick (viscous) then it reacts too slowly and the robot might fall over before it knows it is tilted.

One device used in spacecraft, aircraft, ships, guided missiles, and perhaps other devices, which can give directional as well as angular information, is the well known gyroscope. You can find some without gimbals in toy stores and others with gimbals in high school or university physics labs. The theory of gyroscopes is mathematically explained in physics texts.

Basically the gyroscope is a heavy wheel (much mass) which is caused to spin very fast. The wheel is mounted in a very particular type of suspension system (the cardan suspension) which is a series of rings called gimbals. Each ring is pivoted inside the other such that this permits the fixture to which the spinning wheel is attached—the body or vehicle carrying it—to move up and down, forward or backward, or left and right (translational motion) without disturbing the plane of rotation of the wheel. The pivots in the gimbals permit angular motion around the wheel without exerting pressure on the wheel axle, and you can measure the angles between gimbals to determine changes in direction of the body carrying the unit.

The wheel, when spinning, maintains its plane of rigidity—that of course is what makes it possible to use it as a position sensor. We have written about gyroscopes in our *Aviation Electronics Handbook* (TAB book 631) and in our *Handbook of Marine Electronic and Electrical Systems* (TAB book 939), so you might want to examine these for a more detailed account than we present here. Now, let us examine Fig. 3-9.

Since the rotor of the gyroscope tends to remain in the same plane *when it is spinning fast*, it can be used as a reference. That is, its axle can be used as a reference since the axle will always point in the same direction. Through the gimbal system, the body holding the gyroscope can turn or twist or move in any direction and the various angles through which the body turns will be indicated by a closing or opening of the gimbal angles, when it is so oriented in the body that angular motion will make this happen. Notice, however, that translation or sideways movement or up and down movement or left and right movement of the rotor and gimbals will not close the gimbal angles and will not affect the rotor's "rigidity-in-space," as

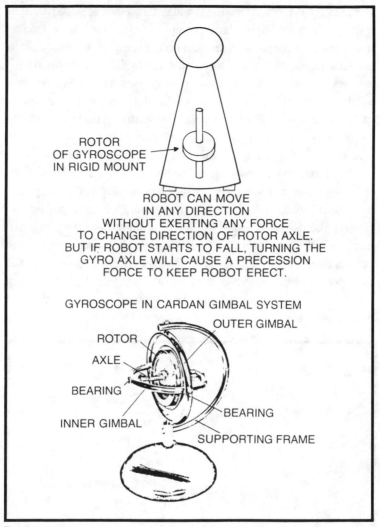

ROTOR
OF GYROSCOPE
IN RIGID MOUNT

ROBOT CAN MOVE
IN ANY DIRECTION
WITHOUT EXERTING ANY FORCE
TO CHANGE DIRECTION OF ROTOR AXLE.
BUT IF ROBOT STARTS TO FALL, TURNING THE
GYRO AXLE WILL CAUSE A PRECESSION
FORCE TO KEEP ROBOT ERECT.

GYROSCOPE IN CARDAN GIMBAL SYSTEM

OUTER GIMBAL

ROTOR

AXLE

BEARING

BEARING

INNER GIMBAL

SUPPORTING FRAME

Fig. 3-9. A basic gyroscope and a rotor positioned to keep a cone-shaped robot upright.

this kind of movement does not tend to try to *turn* the rotor axle in any direction. If you do happen to exert a force on the rotor axle which would tend to turn it, then you will encounter a force from the unit itself which will resist this force and the rotor will, instead of moving as you expect, precess in a direction which is 90 degrees away from the direction of the force application, in the direction of the rotation of the wheel. This phenomenon is called precession.

In airplanes, guided missiles, and space vehicles, small magnets or "torquing motors" are used to apply forces to the rotor axle to keep the axle properly aligned with some distant point in space, such as a far away star. You see, the rotor axle will drift in position with time because of friction in the rotor bearings and perhaps other tiny unbalances which are too small to be detected in manufacture. Sufficient here to realize that a small, but costly, gyroscope might be used to give a robot a sense of the vertical and to tell it when it is bending or leaning in any direction from that vertical. Motor forces can then be put into action to keep the body stabilized during motions of this kind that may occur, or to cause a rebalancing of the robot if it tends to fall. A guided missile is made to hold a vertical position on take-off and does not rotate at all during flight because it is gyro-stabilized. To convert the physical indication of change in angle of the gimbals to electrical signals, potentiometers may be mounted at the pivot points of the gimbals so that movements of them changes a wiper position. Thus you get electrical voltages proportional to the angular gimbal displacements.

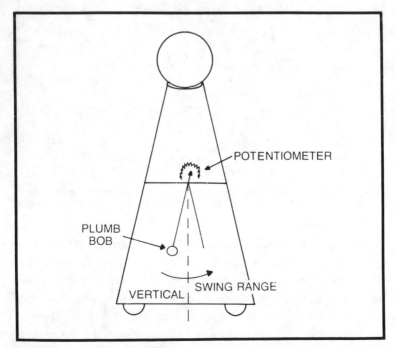

Fig. 3-10. A pendulum plumb-bob sensor.

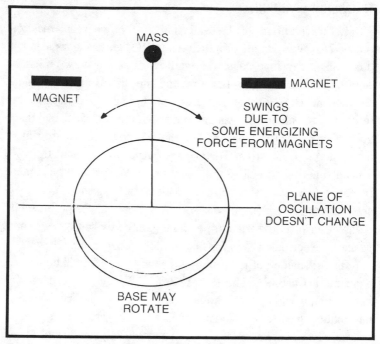

Fig. 3-11. The inverted swinging pendulum.

THE "PLUMB BOB" VERTICAL SENSOR

A surveyor's plumb bob can sense the vertical, but it can go into oscillation if mounted to a moving platform. This is not a good vertical sensor to use in a robot unless it is heavily damped, as was discussed in an earlier chapter. But there is a kind of pendulum, an inverted one, which is caused to swing back and forth by electrical-magnetic energy, and this will tend to hold its plane of oscillation even though its base may be turned. Sucn a device is illustrated schematically in Fig. 3-10.

There are some difficulties with this as a reference. The swing has to be constant and since it might be affected by gravity if its base is moved in an angular direction other than about its center of oscillation, this might not be a unit which has robotics applications either. But the idea could possibly be improved upon and a tiny one be made by someone who likes to experiment. It could be used to keep the robot in vertical position or to give directional reference to the robot. Figure 3-11 shows a general idea.

TELEVISION SCANNING

Let us refer back to the television concept wherein a matrix is scanned to give voltages proportional to the light and dark illumination on light sensitive cells. We want to explain what we mean by scanning. Examine Fig. 3-12. Now imagine that S1 moves to line A and then switch S2 *moves across* from e to h. Only the voltages of the top row of the photocells will be sent down the line of S2, and these voltages in sequence. Now imagine that S1 moves to B and again S2 moves across from e to h. The voltage output will be those from the second row of light cells and in sequence. With this arrangement repeated till S1 goes through D and then starts again at A, you will see that the voltage output will be a series of voltages produced by scanning each cell in turn. Proper timing of the two switches, which would be electronic types for fast action, is necessary.

The resolution of the voltages coming from S2 will be further examined in Chapter 6, The Robot "Brain." We will say here that this information would go to a comparison unit which, in turn, also gets information from a previously programmed grid (or equivalent source). The comparison circuit then has two voltages to look at and based on the accuracy of the two in comparison, some other operations can be initiated in the robot such as steering, slowing down, or stopping. A very simple 2 × 2 matrix using photocells probably could be mechanically scanned for some elementary robot applications such as determining if people are moving in front of it or raising a hand toward it.

TELEVISION 3/D REPRESENTATION

There has been considerable study by scientists of the possibility of using two television cameras, one for each eye of the robot, to give it a three-dimensional vision. That is, it would have depth as well as height and width. An ordinary scanner, such as we've discussed gives a flat two-dimensional vision. It does not really give any real information as to the distance to things in front of it.

In a study by the Jet Propulsion Laboratory, California Institute of Technology at Pasedena, Yakimovsky and Cunningham determined that obtaining 3-D measurements from a stereo pair of TV cameras was a task requiring camera modeling, calibration, and matching of two images of a real 3-D point on the two TV pictures. A

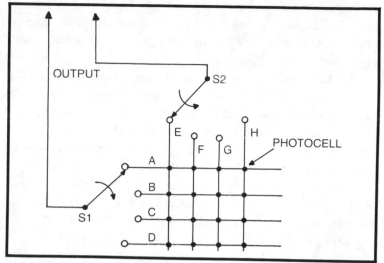

Fig. 3-12. A matrix scanning concept.

model, manual or automatic, which could do this was implemented. This was a real system which was actually operated and actually gave that illusive 3-D effect at a distance of about two meters. Probably the TV cameras have to feed a computer which, in turn, was programmed to compare points. When looking at the same point through two separated eyes it determined through some mathematical magic (triangulation formulas perhaps) the depth or distance to the point. We need 3-D vision as well as coordinated, but not identical arm and hand movements of a robot if it is to become anything near a humanoid level of automation.

Depth, or distance, as we have previously indicated is available through any kind of radar type system ranging unit. It might be a laser, sound, or microwaves. This system can always be used in conjunction with other indicators (sensors) to give some depth and distance information to the robot. But if the TV camera system works completely over the necessary ranges, we have one unit which might do the job, and we would be approaching a real pair of "eyes" for our robot.

THE LASER DANGER

It is necessary, when discussing the possibility of using a laser in a robot, to point out again the danger of using such a device. A laser is an active unit, meaning that it sends out a high intensity beam of light. This can be so strong and bright that it can burn holes in metal strips. We have actually done this, burning holes in razor blades with a ruby laser in the laboratory.

But the danger we want to alert you to here is that looking at the laser beam, even from the side, or in some cases, just the high intensity reflection from the beam from a highly reflective surface can be enough to damage your eyes. Thus the use of lasers in Robotics might not be such a wise step. While it does give a nice tight beam which can give perfect directional and distance control information in the proper circuitry, the danger of its use in a home robot might be precluded because of the hazard potential to the eyes that it presents. Be aware of this possibility.

A SIMPLE LASER

With adequate warning then, and because some modification of the laser might come about which would make its use practical, and because this is a general book and should include all information possible, we now examine the construction of a simple ruby laser.

If you place a ruby rod in a polished ellipsoidal cavity so that it is positioned at one focal point, and you place a high intensity flash tube, electrically operated like your camera strobe, at the other focal point, seal the unit up so that only the tip of the ruby rod has an opening outside, then you have what is essentially a ruby-rod laser. When the flash is ignited by a pulse of high voltage electricity the ruby rod is stimulated into a light emission for just a very small fraction of time and it emits its beam of parallel light rays. These do not diverge as does a normal flashlight beam, the laser beam remains tiny at a distance as far away as the moon.

Now, whether or not a reverse laser will ever be developed is not known. This would be a laser which can act in reverse, receiving the beam of light and turning it into a flash, or perhaps, using some kind of gaseous laser, might turn it into a voltage output. What a nice tight beam for transmission and reception this would be. But whether or not this will come about remains to be seen. Narrow vision optics using lenses have been developed, but these usually have a very finite focal range, and are not usable over long and short distances at the same time. We have sketched the ruby laser concept in Fig. 3-13. At 3-13 (b) we show a light reception concept.

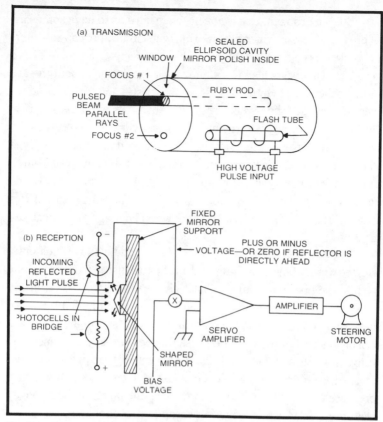

Fig. 3-13. A pulsed laser concept for transmission and reception.

FORCE FEEDBACK SENSORS

When we consider the ability of the robot to grasp objects, we must have some indication, electrically, which can be used to determine how strongly the hand is grasping the object, otherwise it might crush everything it touches! We have mentioned tactile hands. These might be made of a flexible plastic, but a very strong plastic which will not tear or deform when used like the skin of a hand. Inside these hands, of course, have to be the equivalent of bones and tendons, and these would be metal linkages, or possibly very strong plastic linkages which make the fingers work. The fingers might work in unison, that is all together, which makes the linkage very simple, or they might work independently, which means a lot of linkages.

119

But the problem, that is *one* problem here, is to have some kind of pressure sensor which can inform some circuit, perhaps an internal control circuit, how tightly the object is being grasped. We have considered a few such sensors, now we want to think about the use of hydraulics to move the robot's arms, wrists, and fingers, and we want to consider the use of pressure sensors as an integral part of the hydraulic output piston cylinder. These hydraulic systems can be used to give great force, they can be designed as small tube-type units so they may fit into arms etc, and they may be powered from an electrically driven hydraulic pump. Aircraft use a jillion of them as do rockets and much construction machinery. The advantage of this system is speed of motion and *great* power advantage as we shall see later on.

Right now we consider a piston in a cylinder into which pressurized oil may flow into either end and be exhausted from either end so that the piston may move. That gives us a concept of where to make a force measurement. We can imagine, as in Fig. 3-14, that we have such a cylinder and we have affixed a pressure gauge to each half of it as shown. Now, if the pressure in either side increases, the gauge will show it, and will show how much that pressure is increased, (or decreased as oil is exhausted from a side). It will take pressure to cause the piston to move, and the oil has to be removed from the side which it moves toward, or it cannot move at all.

We know it is easy to convert, using a sensitive potentiometer for example, the movement of a dial into an electrical voltage. Thus we have an ideal system here for force feedback type of control. We might use other types of tranducers (devices which convert pressure to electrical signals proportional to the pressure) built right into the cylinders. Now, if the robot's hand grasps an egg, for example, the force feedback signal though an amplifier and into a computer comparator can stop the grasping when the pressure is just sufficient to permit lifting the egg without (hopefully) breaking its shell. If the pressure is monitored to individual fingers they *each* can be made to grasp with a desired force. If the hydraulics operates all fingers against a thumb, then one force feedback will do the job, but some difficulty may be experienced lifting or grasping some shapes. You see, when we use our hands, the fingers curve about and around the objects we pick up to get a good grip and hold firmly against slipping

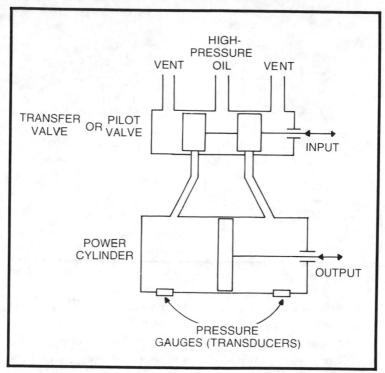

Fig. 3-14. A hydraulic power cylinder with its control pilot, or transfer, valve.

or sliding of the object grasped. A clamp-type hand will probably have slippage because it cannot curl its fingers, but it is simpler to construct. Fingers are better, but the number of fingers moved may not have to be four—as we have. It's an idea that you will want to think about. We explore this idea more later.

CARBON GRANULE FORCE SENSORS (RESISTORS)

One type of sensor which may well adapt to a force measurement is that which is used to convey the pressure waves of sound into electrical signals. Any microphone is a force measuring device if it is used in this capacity. Think of a carbon mike. The carbon mike has a chamber filled with tiny granules which are compressed and released by the action of pressure waves on its diaphram. Why not, then, use this concept to make a force sensing element for a robot's hand. You might use a plastic arrangement in which tiny channels are filled with carbon granules, and electric connections made to each

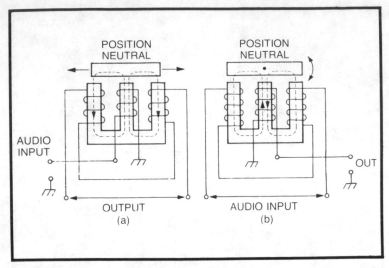

Fig. 3-15. Two basic types of E-core transformers.

end of each channel. As the hand compresses, there will be pressure at various points which will cause the granules to compress and so you can generate feedback signals to be used in controlling the amount of pressure exerted. Other types of microphones might be used in this same capacity, but the carbon-granule type seems the simplest and could be the most reliable for systems where reasonable pressures are to be expected. Of course if the grip is to be very strong, then strain gauges, which can also measure force in the bending of the fingers, or other kinds of elements might be used. These kinds of force measuring sensors are very strong and capable of withstanding large pressures without damage to themselves.

It is also a true fact that in case our robot is required to handle delicate and perhaps hot objects, the sensor must be capable of withstanding temperatures of the range expected and to do this without damage to itself. Sometimes the force sensor might be located away from the hand proper, and the pressure felt through some lever arm as the jaws of the hand grips objects. This way the sensor is removed from the heated object, or other kinds of objects which might damage it. Think about it. We will venture to assume that you will visualize other means of operating sensors that can be perfectly elegant solutions to the force sensors operating requirement.

THE E CORE TRANSFORMER

This sensor has long been known of and used in industry and telemetry for the measurements of physical displacements. It can be used to indicate a small movement, or through a lever system, a large movement of arms, fingers, etc. Or if it is attached to a base in a proper manner, and the base deforms slightly with pressure, it can sense the pressure applied. Its diagram is shown in Fig. 3-15.

An illustrative graph of its output vs. displacement is shown in Fig. 3-16. Notice that the output voltage is very linearly related to the displacement for small excursions, a very desirable characteristic.

To use this transformer you need an AC signal of some kind which is applied to the center windings marked input on the Fig. 3-14(a) and the output windings of (b). This signal may be audio or 400 cycle AC or even 60 cycle AC. This signal will create magnetic fields in the iron outer legs of Fig. 3-14(a) and will cause currents to be induced into their windings. Because these windings are exactly equal in turns, and because the direction of the flux, when the armature is centered, is exactly equal in each leg, the voltage induced into the leg windings will cancel, and the output voltage can be zero when the armature is at neutral.

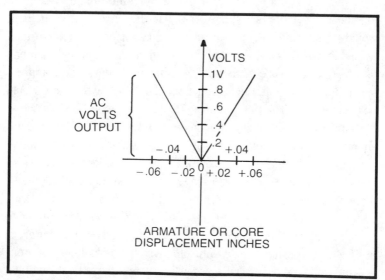

Fig. 3-16. Output volts vs. displacement.

123

When, in case (a) the armature is moved a fraction of an inch from the neutral position, more flux will be induced in one outer leg winding than in the other and so the flux will not cause equal currents. There will be an output which is the difference of the two induced voltages and which has a phase relationship to the input which is that of the larger induced voltage. The phase will change for an equal displacement of the armature in the opposite direction.

Since phase then will be different in the output voltage, depending upon which leg gets the most magnetic flux, we now have an indication, by phase, of which direction the armature moved. Of course we must then arrange some kind of phase comparison circuit to be fed by the reference (input) voltage and the output voltage to give us a usable signal whose polarity is related to the in-phase or out-of-phase condition. But such circuits are well known and proved. They produce this dc voltage output whose magnitude is proportional to the difference in leg signals being compared, and whose polarity, plus or minus, is directly related to the phase relationship of the two signals, input and output.

In Fig. 3-15(b) a slightly different arrangement is used to increase the flux in one leg or the other. A pivoted armature is used here, and thus this becomes adaptable as a sensor for small angular movements (remember the gyro gimbal system?). Figure 3-15(a) is for linear movements. Notice also that the angular arrangement would be suitable for measuring pressures if one end were connected to, say, the inside of a finger to a small plate and the other end of the armature were connected to suitable spring of the compression type. Thus when the finger grasped something the plate would be pushed a small amount, causing the spring to compress. Then at the same time a signal would be generated which would indicate the amount of compression of the spring. Over small regions of motion you can get springs which are quite linear, and since the E-core also has a fairly linear response, you have a good system to use for finger force measurement. The strength of the spring governs the grip.

It is of interest to note that various manufacturers probably design their variable inductance E-core transformers into different packages. You can probably find some which look like small motors or synchros, and others will be small rectangular packages with an arm extending for use with translational movements. You can write transformer or instrumentation manufacturers for more information. Telemetry equipment manufacturers use them a lot and they are a good source of information on obtainability of such units. You might, if you are inclined, wind your own, using a small power-transformer core or the secondary winding of such a transformer and adding a winding to get an E-core like that of Fig. 3-14(a). It is a thought. But remember that these, as robot sensors, will have a wide range of applications, and you will no doubt quickly think of ways and places to use them on your robot.

PNEUMATIC SENSORS

Pneumatic sensors might also be used with an air-power system. They might also be used, some of them, with hydraulic systems. We call to mind that the particular advantage of using hydraulics (oil) as a working fluid over that of using compressed air, is that air is compressible. That means that if, for example, you extend an arm with an air piston and you put enough weight on the arm, the arm will droop slightly as the air compresses, unless the system compensates by furnishing more air pressure. With hydraulics you do not have this situation. Oil is not compressible, and so the arm is essentially locked into position once you have moved it where you want it.

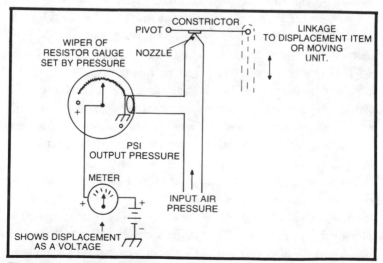

Fig. 3-17. A pneumatic displacement sensor.

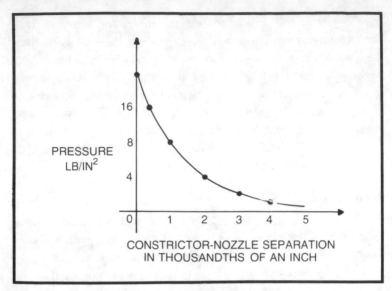

Fig. 3-18. A relative pneumatic operating curve.

But we need to examine at least one type sensor related to the use of air as the working medium because it may have some application. Some people say that because the mass of air is less than oil, an air system operates faster than a hydraulic system. Perhaps this is true, but in our visualized applications in robots, that kind of speed normally wouldn't be required. Examine Fig. 3-17 and 3-18.

Now we can see here that air is applied at constant pressure to a constriction which can be varied by the movement of a linkage. There is a pressure gauge designed as an electronic sensor or it could be some other pressure-sensitive unit, perhaps one plate of a variable capacitor which bends with pressure applied and thus changes the frequency of an oscillator circuit attached to it. The output nozzle exhausts the air into the atmosphere against the constriction; and the constriction may take other forms, such as a screw in an output hole.

Now, when the constrictor is far away and the exhaust nozzle exhausts the maximum amount of air, there is little pressure on the pressure sensor. When the constrictor is moved to close-down the output exhaust, pressure rises to the maximum when the exhaust is fully closed. Thus we do have a physical movement being expressed as a pressure, and through a suitable transducer which converts this pressure into a proportional electrical signal, to a voltage output proportional to the constrictor movement.

This voltage can be used in various ways in a robot. It can indicate the movement of arm, finger, head, or eye. If the system is pneumatically powered, it will always be a feedback signal which opposes a command input signal, and thus will show by comparison when a desired movement has been accomplished.

The system requires a compressed-air source of power which might come from a small air pump or from a compressed air bottle. Like the hydraulic system, which also requires a pressurized drive source, this can mean extra weight and bulk inside the robot. But then, the electrical system will require batteries and these can also take up space and be bulky. So the system you use, electrical, hydraulic, pneumatic, or whatever combinations of these systems you choose, might be in terms of what your robot must do, how well it must do it, and how fast and strong he has to be to do it.

COLOR SENSORS

Robots which can detect color (chapter 1) will use some kind of photocell which is sensitive to the radiations in the various bands of light. One such cell is the Clairex CL5M5L which is often used in color analyzers. It is used in conjection with a color filter lens. One source for more information on how color analyzers are used can be found in such publications as Kodak's *Printing Color Negatives*. What happens in a color analyzer is this: the amount of light on the cell varies, depending on whether you use a red, green, or blue filter between the cell and the light source. Because the light intensity changes, you will find a change in current from the light cell, which can be identified in a small computing circuit as calling for a certain response by the robot when that particular light is present. You will get, say, a maximum current from red with a certain filter, and that current may activate a relay. Thus your robot responds to colors represented by this kind of light. You might want to experiment in this area if you want your robot to detect and announce the colors you place in front of it, or, perhaps, to follow a path on the floor which is defined by some particular color other than white. There are lots of possibilities for this kind of sensor in robotics.

A SENSOR FOR SMOKE

Although this kind of sensor is quite common nowadays in home fire alert systems, we feel they may also have an application in a robot. They might be used as a fire alert part of the robot system. That is, if the robot smells smoke, it can warn the occupants and since the robot can be programmed to walk around the house every few hours as a security agent, this system can sample many areas well. But there is also the possibility that detection of other gaseous substances might be made through this kind of sensor when properly adapted for the other gasses, so let's examine the smoke sensor in more detail.

It is most interesting that this type of sensor uses a harmless ionized ejection of helium atoms as its main ingredient. These come from an element called americium which is put into two separate and very small chambers. One chamber is open to the atmosphere, the other is sealed and used as a reference. There is a detector for the ionized atoms which converts the number of them per second into a tiny electrical signal, much as a geiger counter would do when feeding the impulse rate of sampling of atomic particles to its indicating meter. Now one detector is in the sealed chamber and one detector is in the chamber open to the atmosphere.

In the smoke detection application, when a small bit of smoke enters the open chamber, it causes a change in the number of charged particles and thus a change in the voltage detected by the tiny transducer contained therein. When the voltage output of this detector is then compared to the reference detectors voltage output through an integrated circuit, the change, or voltage difference, when large enough, will set off an audible alarm.

It is said that if the charge drops one-millionth of a percent the integrated circuit computer-comparer circuit will sense this change and set off the alarm. It is also claimed by manufacturers that these units can differentiate between cigarette smoke and smoke due to a fire. This is probably because of the kind of voltage change that occurs with each type of smoke.

This is also something that makes the unit attractive as a robot sensor. Perhaps with some experimentation and computer adjustment and voltage comparison techniques, the sensor could be used to detect other types of gaseous emissions, and thus provide our robot with a nose.

The radioactive substance americium emits alpha particles of the harmless ionized helium atoms which charge the air in each of the chambers. Thus an electrostatic type of measurement is possible, that is, the charge or change of charge between capacitor plates. A capacitor is charged when it has an electrostatic field between its plates, as you will recall from basic electronics theory. The charge quantity can be proportional to the size in farads of the capacitor; thus, with a good charge we will have a measurable quantity to detect through the detectors. You will also recall from basic physics that the alpha particles travel only a short distance. It is the beta particles which are heavier and which can travel far enough to be dangerous. We presume that there are no Beta particles emitted from this element, or if they are, they are so few as to be inconsequential.

Thus we find another off-the-shelf item which we can incorporate into our robot. We can use it as it is, or modifications, depending on the application to which the robot is to be put. For more information on these units, write JS&A Sales Group, Dept. WJ, One JS&A Plaza, Northbrook, Ill 60062.

A SPEAKING-VOICE SENSOR

We might want to talk to our robot, and we might want it to talk back. The sensor for the human voice is a microphone, which might be any of the several commercial models, a small crystal microphone, for example. The robot's ability to speak back might be accomplished through an artificial larynx, or voice box. This device would, of course, be computer operated.

Now we would assume that the robot has a voice detection system incorporated in it so that it can identify your voice and will be able to recognize several words. MIT has done much work in voice recognition systems over the past few decades and has proved that this is possible. But then, having identified the key word or "words" spoken to it, an integral program inside the robot, activated by its computer brain could cause it to speak certain preprogrammed sentences or words back at you.

We consider the use of an artificial larynx because it is another off-the-shelf item. It may be found in medical supply and electronic houses. It was developed in several forms to be used in humans who have, through disease or accident, lost their ability to form sounds associated with the human voice system. As you know, we humans have vocal cords which are stretched and loosened by neck muscles to make the tones we emit. The air rushing past these vibrating membranes makes the sound.

So we might obtain such a unit, an artificial larynx, which will provide some kind of acceptable human voice sound. Next we must try to have a word forming section and that might be more difficult, but maybe possible. As you know we form words by the relationship of our tongue, teeth, cheeks, lips, and mouth cavity. We produce a sound in our throat and then "flip it around" in our mouth so it comes out as a word sound. We say these sounds are words simply because our minds have been taught this, otherwise they would simply seem to the hearing as a variable change in pitch, timbre, sharpness, duration, and sibilance of sound, and entirely unintelligible. Think of your reaction when listening to foreign language you do not understand.

Form some words and see how your mouth positions all its parts to make that word. Then you may understand this concept. Difficult to instrument? We didn't say it would be easy, but is seems feasible. But not by having tiny computers activate programs to form mouth, teeth, and lips positions while a tone is blown through an equivalent mouth. Instead, breaking down the words into basic sounds, then recreating those sounds in a computer in the way that a human does gives us what is called a Komputer Talker. Well, at least a robot won't be telling us off because we won't program that response into it. But as yet, the humanoid, cybert, or cyborg cannot respond to sounds which they hear unless it is a preprogrammed response.

Some commercial robots now in existence have built into them a playback unit which has a large track of various responses to certain sounds (words) which it hears. When asked, for example, "How are you?" a little cassette inside the robot spins madly to the appropriate answer *to that sound*, and when it finally succeeds, it stops and plays back the answer, "Fine. How are you?" The sound would radiate from a tiny speaker in the mouth chamber. That is how it is generally done as of this writing. Later, perhaps, it will be done in the manner previously described, wherein the robot actually simulates human speech and perhaps even "thinks" of possible things to say.

THE SPEECH PROCESSOR

Let's examine an item described in a newspaper article. Speech scientist Kathy Fons picked up a small gray box and punched a few buttons on the side of it. "I can talk. I am using an artificial voice," said the machine. It was true; without tapes or cassettes or recordings of any kind, the box produced recognizable human speech.

The manufacturer, a division of the American Hospital Supply Corporation of Evanston, Ill. claims that this is the first artificial voice machine available for speech-impaired people. It was first marketed in 1978 in two models, one 8 by 14 inches, the other 5 by 8 inches. Cost was about $2,000. The company's HC Electronics Division is the research organization developing the unit.

Now this machine is said to produce a low pitched male sounding tone with a slight Scottish accent—if you can imagine that—and it originates its speech with a small computer. It can say about 1000 words, phrases, or programmed words which have been placed into it previously in the computer memory. When buttons are pushed, the order in which they are pushed selects words which then come out audibly in the button-pushing sequence. Thus a kind of speech is accomplished.

It is said that although the machine speaks slowly its sounds are easily understood. It is programmed phonetically, and is designed for deaf-mute persons and others who cannot speak. HC Electronics has said the unit will be available only to medical specialists and speech therapists at present, but it is a forward step in communications with persons who have been stricken in some manner or born without the ability to speak. They must have the use of at least fingers and have some degree of intelligence to operate the unit. (Doesn't that meet our requirements for a robot?) There are simple units for persons who have little ability manually or mentally.

What all this means to us—aside from the marvel of our medical technology—is that we might now visualize a robot which does not use recording tapes at all. By being able to recognize words spoken to it in sentences—another area of intense development nowadays with some accomplishments—it can search out words and word patterns that would make up a suitable response. If you want to boggle your mind a little, try thinking out how a computer would select the *proper* group of words from say 1000 available to answer the question "How do you think?"

Well, we might listen to voice sounds and then investigate the phonetic possibility of using oscillators with just the right combination of tones to represent a male (or female) voice in just one intonation—a kind of monotone—and then by clipping, shaping, and such other electronic management modify that sound so that it comes out as a form of phonetic speech. You will have the real basis then for a talking robot. Think how interesting this could be. Wouldn't you like to have an *unprogrammed* conversation with a machine. Wonder what it would tell us?

But enough here of sensors and such things. We must move on and we need now to explore some power sources for robots, for if we don't have them, the robot will just sit (or stand) there and do nothing! Kryton says he likes this next chapter.

CHAPTER 4
PRIMARY POWER SOURCES

We come now to an investigation of primary power sources for a robot or, to think ahead, for an android. Most obvious are electric batteries. We need to examine several types of these and also some pneumatic and hydraulic power sources. Then we'll glimpse into the future to atomic power sources. This latter type, if well contained and harmless to humans, would be ideal for they would rarely have to be refueled.

BATTERIES

Let's consider some batteries which might be used. Dry cells are out. They have too short a life and are not rechargeable. But don't mistake them with their small counterparts, the nickel cadmium batteries which look like dry cells. Nickel cadmium batteries *can* be recharged, as we shall see.

The first battery to be considered is the regular car battery shown in Fig. 4-1.

This is an electric automobile which we are told is being manufactured in larger and larger quantities. The battery used is the lead-acid type. A second type battery, which is used primarily in aircraft, is the nickel cadmium. It may have some advantages over the lead-acid type even though the former is the most popular and the most economical. A third type is the lithium sulphur battery, which is now under development and test.

The Lead-Acid Battery

Of the two types of batteries which can be used to power your robot the lead-acid battery, the most common, is the kind used in your automobile. It may be sealed or nonsealed. If it is sealed you do not have to add water, but you cannot measure its charge with a hydrometer either. These are almost spill proof although some types will leak electrolyte if turned up at a large angle.

The lead-acid battery might be damaged if it is discharged to too low a level or charged too fast or overcharged. If it is charged in reverse it is finished, and probably the charger will be damaged also. Care must be taken never to attempt to charge this battery backward. When it is charged or discharged at too great a rate there is a heating of the plates. This causes them to shed some plate material which then causes a short circuit in the cell where this builds up. Then the battery is ruined. For maximum battery life you must stay within the charge and discharge limits set by the manufacturer.

It is true that in the discharge-charge cycle the plates tend to loose some material so that eventually they will not take or hold a charge. Internal electrical leakage will take place in time. Have you ever noticed how closely the failure of a lead-acid cell follows the expiration data on the guarantee? Maybe its a coincidence. Anyway, when you are running motors there is initially a large drain to start the motors. Usually the battery has to furnish a smaller current after the motor starts running. Here are some tips to get the best use from this kind of battery.

- New batteries should always be fully charged before use.
- The life expectancy of your batteries can be increased by keeping them fully charged.
- Keep water to the proper level (if the battery is not sealed). Use *distilled* water.
- Never discharge your battery by over use. This shortens the life of the battery.
- If possible measure the charge with a hydrometer. When glass tube is filled the number on the float at the water level is the specific gravity reading. Car parts stores have hydrometers.

And here are some specific gravity readings which might be useful:

Fig. 4-1. Electric car batteries.

- Good charge = specific gravity 1.230 to 1.295
- Requires charging = 1.230 or less per cell
- If you get a reading *above* 1.295, your cell was improperly activated when new. Your battery will have a short lifespan, and will give poor operation.

Of course this kind of battery—any kind, in fact—should not be exposed to handling by any persons. An acid-proof box of plastic or other suitable material should enclose the batteries in such a fashion that no liquid can leak out, even of sealed batteries. Examine Fig. 4-2 for an example of a good battery container.

The Nickel Cadmium Battery

On aircraft nowadays the standard cell is the nickel-cadmium battery, which has a long life and can even be charged in reverse without harm. It can be completely discharged and will "come back" when properly charged. Thus it seems to have some advantages in life and ruggedness over the lead-acid battery. There are some differences between these two batteries, however. The nickel-cadmium type generally has a lower cell voltage than the lead-acid battery—*1.35 volts per cell* as compared to *2.0 volts per cell* for the lead-acid type. The lead-acid type has a gradual discharge curve to failure while the nickel cadmium has an almost flat discharge curve until it is exhausted, and then it has a very abrupt dropping of output voltage and current. You must be careful not to overcharge nickel-cadmium batteries or to charge them too fast. Note the curves of Fig. 4-3.

Of course one must be aware that hobby supply houses have permanently sealed nickel-cadmium batteries which can be re-charged many many times and which supply a good output for various uses within their capacity. For some robot applications you might want to investigate these. But remember that the hobby houses normally do not have the high capacity cadmium cells for powering drive motors. These you may want to obtain from a manufacturer such as Marathon Battery Co. P.O. Box 8233, Waco, Texas 76710. Write to them if you have a special size, current, or voltage requirement which must be met in your robot.

136

Fig. 4-2. A neat and safe battery container.

BATTERY AMPERE-HOURS

Battery output is stated in terms of ampere-hours or possibly in some applications as volt-amps which means watts. A typical battery for a car may have from 50 to 600 amp-hours of use, meaning that it should, under good conditions and when fully charged, be able to deliver, say, 50 amperes for one hour, or 1 amp for 50 hours. We need to have some idea of the current drain in amperes that our complete robot will require in order to specify the best size battery to use in it. We also must be sure we *do* have enough amp-hours to permit everything to work as it should over the time period we desire without recharging.

When the demand is large—for example when a wheel chair must be used all day, perhaps by a student attending university

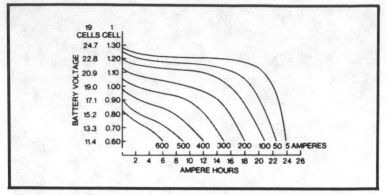

Fig. 4-3. Discharge curves for a nickel-cadmium battery. These are curves for a Marathon CA-21H-1 battery at room temperature.

classes—and recharging is only possible once a day, you can increase ampere capacity by paralleling two 12-volt batteries.

The Charge-Discharge Cycle

Normally, the auto battery we are familiar with is charged by an alternator or generator driven by the car engine when it runs. This prevents a discharge depth of the battery below, perhaps, half-discharge value. Of course when the car is running, the electrical equipment receives power from the generator or alternator as it "floats" across the battery. So the battery doesn't run down unless it is damaged or defective.

With a robot we have a different situation. The charging is accomplished only at specified times. The batteries must be capable of running the machine until a charge time arrives. This will probably cause a deep, or near-complete, discharge of the battery. With the lead-acid cell this deep discharge may produce a near zero voltage output. With the nickel-cadmium cell it results in the 1.2-volt level, the level for nearly complete discharge of this kind of battery.

With nickel-cadmium batteries a charge-discharge condition different than other batteries must be understood. Unfortunately this type of battery has a "memory." It tends to remember how much it is discharged under normal conditions and soon won't hold a charge beyond that. Woe is us! Of all things, who would expect to have a battery remember. So, what we have to do with this battery is give it a deep discharge and a full charge in a series of cycles to keep it happy. It is said that three such deep-discharge and full-charge cycles are needed to remove the memory capability from a battery. This must be done every once in a while whenever it is beginning to remember again.

Those happy people who build and fly radio controlled model airplanes and operate radio controlled cars and boats (a plug here for our *Flying Model Airplanes and Helicopters by Radio Control*, TAB book 825) have learned of this memory problem. They follow cycling rules to insure that their batteries won't develop a memory to plague them in their flying, racing, or sailing operations. We must take a tip from them and, if using nickel-cadmium batteries, be sure to discharge the batteries in normal operation to a completely discharged level; then charge them fully and repeat this process at least three times every once in a while to eliminate memory problems.

Now, the deep cycling situation, which means extended use and then long recharge times (or relatively fast discharge times) presents a different kind of problem for us when we use lead-acid cells. This battery may not be able to survive such treatment. In a car it doesn't normally have a deep discharge situation and thus does not require a long, hard charging effort. Remember the car battery always works with nearly a full charge on it.

So what we need then is a specially designed battery to handle the deep discharge and relatively fast charge rate over and over again without causing battery deterioration. One such lead-acid battery is said to be manufactured by Gould Co., 10 Gould Center, Rolling Meadows, Ill. 60008. It might be worthwhile to contact them, or other good battery manufacturers and find out about deep cycling lead-acid batteries, if you want to use this type in your robot. Gould makes batteries for such applications as marine trolling, running accessory motors, and powering recreational vehicles, where demands are large and recharge is accomplished with relatively high currents to reduce the charging time.

The Charging Process

Every battery has an optimum charging rate. The manufacturer normally supplies this with the battery when it is purchased. You need to know this. It may be that you must *never* charge your battery at a rate over 20 amperes, for example, and if you do not know this and just think of putting your battery on a high rate, say 50 amperes, to permit a charge in two hours instead of five hours, then you might wind up with a damaged battery. Batteries cost money and so we don't want to ruin them in this manner or injure them needlessly.

Normally, when charging a battery you follow simple rules:

- Remove all load from the battery.
- Check the charger to be sure you are charging within the recommended rate.
- Be sure your battery is kept filled with the proper electrolyte (distilled water for lead-acid cells) if they are low.
- Keep the connections to the battery terminals clean and tight.
- Charge the battery the required time and provide good ventilation in the charging room or charging location.
- Be sure the charger has the capacity to charge the batteries in series or parallel if you use them and charge them this way. Or be sure to disconnect the batteries so they can be charged individually, then reconnected in series or parallel, however they are normally used in the robot.
- Be sure the charger is fused so it will not burn up if batteries happen to fail or short circuit, or are defective when put on charge.
- If possible get a charger with an accidental battery reverse-connection protection feature in it. This type will sense if you get the battery connected incorrectly and will not charge when this condition exists.
- Of course, always be sure to connect the battery positive terminal to the positive terminal of the charger, and the negative terminal to the negative terminal (lead) of the charger. It is best to always connect directly to the battery and not through the robot frame or through a small wire which happens to go to the battery.
- If any acid spills (lead-acid type) or sweats out under charge, then use baking soda to neutralize the spill immediately. Do not try just to wash away with water.

- Keep sparks away from charging areas. Any secondary type battery may generate an explosive gas which may somehow be ignited by sparks, flame, or cigarettes.
- Wear rubber gloves when handling batteries to protect against accidental spillage or sweating out of the electrolyte (even of so called sealed batteries) as this spillage can cause severe burns to skin.

The "C" Factor

With nickel-cadmium batteries there is a C (capacity) factor which must be considered when charging. This is the amount of current the battery can discharge in one hour. You will probably associate the "C" factor with the amp-hour rating of a battery. Thus if you have a battery rated at 100 ampere hours, it will deliver its rated capacity in C/4 hours if used at the rate of 25 amperes discharge. Here C is the 100 amp-hour capacity.

Now, with nickel-cadmium batteries there are several methods of charging, and the rate of charging is related to the capacity C of the battery as follows:

- Trickle charge is at about 100/C for 100 hours.
- Slow "overnight" rate is at C/10 for 10 hours.
- Quick charge is at C/2 for 2 hours.
- Fast rate would normally be at C/1 for one hour.

Of course you must always charge a battery longer than you plan to discharge it. So, if you want to discharge for 10 hours you would probably charge for a much longer time, perhaps a 24-hour period. You would also probably use a constant current charging rate of from 1.4 volts per cell to 1.6 volts per cell for the nickel cadmium battery. It is interesting to examine now the curves of Fig. 4-4 for some direct relationships concerning the nickel-cadmium battery.

Overcharging A Multicell Battery or Multiple Batteries

If you have a multicell battery (or multiple batteries) in your robot system, and you'll probably have this arrangement, you can discharge the battery completely and then find yourself in trouble. What can happen with nickel-cadmium batteries of the multicell type is that the cells are not always uniform and thus one may go dead while others are still operating and have some life in them. Then when you keep using the battery you may get a reverse charge put on the low cell by the other cells. This can cause damage to the low cell if this happens repeatedly.

It is recommended by General Electric that the terminal voltage in discharge never be less than the application voltage per cell multiplied by (n-1), where n is the number of cells, and then subtract the value 0.2

$$\text{Discharge voltage} \geqq (\text{cell voltage} \times (n-1)) - 0.2$$

Thus for a six-cell battery where the cell voltage is, say, 1.3 volts per cell the battery should be recharged when the voltage is measured under load (in normal operating conditions) to be:

$$\text{Recharge voltage level} (= 1.3 \times 5) - 0.2$$
$$= 6.3 \text{ volts}$$

Notice that the full-charge voltage would be for the six cells, or 7.8 volts, so the charging should begin when the voltage drops 1.5 volts from the full-charge level. This means you need to watch and monitor the voltage under load when using a nickel-cadmium multicell battery. The time to recharge a lead-acid battery can be determined with a hydrometer as we have mentioned and will examine in another page or two.

Tapping a Multicell Battery or Multiple Batteries

This may be an important consideration. If, in your robot, you have need for both 12 and 24 volts and you use a part of your 24-volt battery to get the 12-volt level, and then charge the whole battery, you may find an uneven charge situation in the cells making up the 24 volt battery. It may be that the 12-volt half is more completely discharged than the total cell make-up. This could mean a reversal of charge in nickel-cadmium types and a shortened life for part of the cells in the lead-acid type. It is probably better to use separate batteries (or charge them separately) for the various voltage levels you require, and then charge each at its own required rate and for the time each requires. This will insure the best and longest life from the total battery supply.

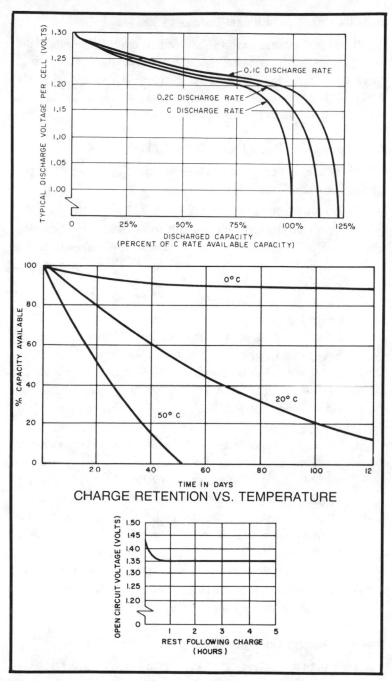

Fig. 4-4. Nickel-cadmium battery curves. Courtesy General Electric.

Best Charging Rate of Nickel-Cadmium Batteries

It is important that you do not exceed the overcharge rate of the battery. For standard cells the charge capability is the 0.1 C charging rate and on quick charge the charge current capability is generally 0.25 to 0.3 C. If you do not exceed these rates you will get the longest life from nickel cadmium batteries.

Precautions When Using Nickel-Cadmium Batteries

These are some precautions recommended by General Electric for nickel-cadmium batteries.

- Do not dispose of batteries in a fire.
- Do not attempt to solder directly to a sealed cell because the cell can be damaged by too much heat. If you need to solder connections buy a cell which has solder tabs put on it when it is made.
- Never lay an uninsulated cell or cells on a metal table or bench top. This may short-circuit the cells.
- Do not wear rings, without gloves, when handling batteries. Severe burns have resulted from short circuits caused by metal watch bands and rings on cell terminals or leads.
- In case you get this kind of electrolyte on your hands wash quickly and neutralize the electrolyte with vinegar.
- Always charge in a ventilated area.
- Since the electrolyte (KOH) may cause deterioration of some metals, such as aluminum or copper alloys, and you might be using these materials in your robot for lightness of weight, strength, and conductivity, be aware that in the vicinity of the battery the metals should be plated with nickel if possible or with some other protective coating to prevent this deterioration. If you have a good protective case, as we have shown previously, of heavy plastic which can withstand the KOH effects you will have provided protection against the electrolyte damage.

Lead-Acid Battery Testing

You can test your lead-acid battery using a hydrometer if you do not add water first. Always test the cells just as they are before charging or refilling with distilled water. Now, when using a hydrometer—a good commercial one, and not the small ball type found in most parts houses for a dollar or so—you should find the following type readings.

- Hydrometer reading from 1.230 to 1.295 at 80 degrees fahrenheit show the battery charged and in good condition. Check each cell for this level.
- If cell is less than 1.230 specific gravity when charged there is trouble in the cell or the electrolyte is not of proper strength. If it does not come up to at least 1.230 when charged a long time, get a new battery.
- If the reading is less then 1.230 put the battery on charge and reach the first two conditions.
- If the battery electrolyte shows a reading above 1.295 specific gravity, then the cell was improperly filled or activated and you can expect short life and poor operation.

Since new lead-acid batteries are dry-charged when the electrolyte is added it is almost impossible to have that same condition exist for refilling with new electrolyte if you seem to have an improper hydrometer reading. During normal operation, if the battery is not sealed (and sometimes if it is the sealed type), you will have an evaporation of the water from the electrolyte and you must replace this with *distilled* water to have good service and long life. Do not try to replace or add sulphuric acid. This is dangerous and you cannot reactivate the battery in this manner. We are really against this. A new battery, while a little costly, is safer and will prove to be more satisfactory.

Sealed batteries of course are not supposed to require water. Some are guaranteed for the life of the car they are used in, and so should have a pretty good life in robot applications. Just keep them charged. Some have built-in indicators, such as a green dot in Delco batteries which is to tell you (when it is bright green) that the battery is fully charged. Other types may have different or no indicators at all, and since you cannot check them with a hydrometer, you never know for sure just what the charge status is on this kind of battery.

What you need for a sealed battery is a voltmeter with a big resistance load across its terminals. There is such a device, usually found at battery shops. It consists of a handle with a load resistance and meter across two prongs. You touch the prongs firmly against the battery terminals and the meter will read the voltage while the battery is delivering a *heavy* surge of current through the resistance load. This will give you some indication as to the batteries overall condition. You cannot check each cell this way because you do not have access to each cell's terminals. But if there is a large, significant drop in voltage when using this load tester, the battery is either not charged or won't hold a charge.

It is true that sealed batteries are best for cars where the voltage regulator in the system keeps it charged to the proper level. If you have some equivalent charger, that is, one which "tapers" the charge as the battery reaches the fully charged state, and if you make it a practice to keep the battery on charge whenever it is not in use, you can be reasonably assured that it will always be ready to work for you.

ALTERNATIVES TO BATTERIES AND MOTORS

We need now to examine some other power systems, the hydraulic and pneumatic (air) systems. These are used more on commercial robots than on types which might be built by hobbyists. It is true that electric motors will always have the majority of tasks to fulfill in a robot system, but since hydraulics gives move power and sometimes smoother action, and air systems are faster and might have some other particular type of advantage, we had better examine them to be sure our knowledge is complete.

THE HYDRAULIC SYSTEM: SOME PRINCIPLES

We have already stated that the use of hydraulics is important because of its power, speed of response, and the fact that oil is *not* compressible. This means that an arm which is positioned by a hydraulic piston in a cylinder can be considered "locked" when it is motionless, held there by oil pressured in the cylinder. Then, too, hydraulic systems are essentially linear, and thus can be adapted easily to arm and even finger and wrist movements using small diameter cylinders and pistons. Actually, however, most wrist movements are made with electric motors, so we find combination systems much in use.

You have hydraulic equipment in your automobile, the braking and steering systems. You know how smoothly these work to give you added power to steer and to apply your brakes. Another example is at the filling station, where they use a lift to raise your car. Air-pressure is introduced to a piston in a cylinder through a valve control at perhaps 200 lb per square inch. Now if the piston area is, say, 144 square inches, then an upward pressure of 28,800 lbs is gained to raise your car. Since a normal car weighs about 4000 lbs, you can see that it will be raised easily, but slowly. Look at Fig. 4-5 for a rough sketch of how the automobile hydraulic system for brakes appears.

Pascal's Law

Let's go a little further into the hydraulic principle, which is basically an application of Pascal's Law. Blaise Pascal was a French scientist and philosopher (1623-62) who did much experimentation with fluids. He said:

"External pressure exerted on a confined liquid is transferred undiminished to all surfaces in contact with that liquid"

This means that an increased pressure applied to any part of a *confined* liquid causes an equal pressure increase throughout this liquid. In an equation form, then, we can state this for hydraulics

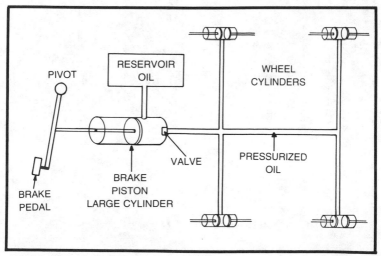

Fig. 4-5. The hydraulic brake system of a car.

where two pistons are used, one to apply the pressure and one to receive the pressure. The area of the two pistons being unequal:

$$\frac{\text{Force A to piston A}}{\text{Force B to piston B}} = \frac{\text{Area piston A}}{\text{Area piston B}}$$

This tells us that if we apply a force of 10 lbs to a piston of area of 1 inch^2 and transmit this force through a hydraulic fluid to a piston of area of 10 inches^2, the resultant force applied to the second piston will be 100 lbs. Examine Fig. 4-6.

What's the catch? Well there really is none. But to satisfy your mind, a further examination of the situation and equation shows that if you move piston (A), with area 1 inch^2, a distance of 10 inches, then piston (B), with area 10 inches^2, will move only 1 inch. You do not get the same amount of travel for the two unequal size pistons, but you do get the pressure, or force, increase, and that is what is important to us here. Be aware that the reverse situation may also be important. If you must move a piston a long way for a small excursion of

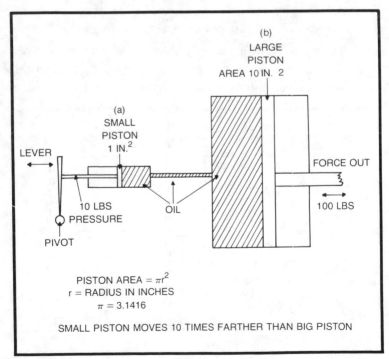

Fig. 4-6. Pascal's law.

some input, then reversing the cylinders will give a 10-inch output for a 1-inch input, but the pressure will be less in the output than the input. This may not pose a problem if the input has plenty of pressure on it.

Now we need to state that if the oil flow is supplied from some pump or other kind of reservoir so that its supply is unlimited (always present), then we need not consider the amount of movement of piston (A) of Fig. 4-6. Instead we will be concerned only with the pressure applied through some control valve to the oil and thence to piston (B).

We have stated that the incompressibility of oil is important, and that is true. If the oil could be compressed then there would be a certain amount of energy lost in compressing the oil to the tightness required to make it *seem* incompressible to any load that might be fastened to piston (B). But hydraulic oil is not compressible and so there is no problem of this kind *when* the system has been purged of air bubbles and the oil is kept pure and is not diluted with some other liquid.

Hydraulic Pistons

We will now examine two basic types of pistons. The spring-loaded type is used in some applications such as to adjust boat trim vanes. The two-port piston, which has a transfer valve attached to it, would be the type probably most found in robot applications. A sketch of each type is shown in Fig. 4-7. Here the hydraulic transfer valve is much simplified, but we will examine it in detail a bit later on.

In the figure we see cylinder (A) in which a piston moves to a forward position, compressing the spring. The spring can push the piston back against the oil if it is allowed to be exhausted from the chamber. The distance it moves will be governed by how much oil is expelled.

Now, to move the piston back in the opposite direction, oil is introduced under pressure, which forces the piston against the spring and both move back until the spring is completely compressed and can move no farther. The amount the piston moves in this direction depends on the amount of oil that is introduced into the chamber.

In this simplest case we see that in one direction it is only the spring force which causes the load attached to the piston shaft to

149

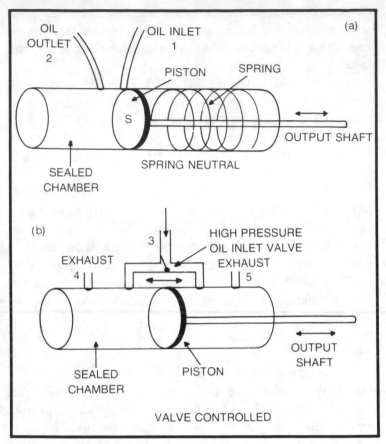

Fig. 4-7. Basic hydraulic pistons.

move. This can be less than the force that the oil pressure exerts against the spring and load in the opposite direction. Let's examine the force situation. If we assume that the oil is put into the cylinder at a pressure of 100 lbs per inch^2, and if the area of the piston is 6 in.^{2}, the pressure on the piston will be 600 lbs. The spring force, however, may vary and reach its maximum in the fully compressed position at 500 lbs. Of course it may be possible, over short movements of the piston, to have a nearly linear spring force applied, and through some linkage, get the maximum actual movement of the load which is required for that application.

But there are possible disadvantages to the spring return piston. It has only the power of the spring on the return, as we have stated. And in some cases, if the oil is light weight (thin), the piston may tend to oscillate a little because springs are natural vibratory elements. This could be true in a fast movement, particularly. But the advantage, again, is its simplicity.

In (B) of Fig. 4-7 you see a piston which is so ported that oil can flow into each end, and can be exhausted from each end. Valves are used, of course, to control the input and output. Actually, when a control valve is used, only one port is required to apply and exhaust oil from each end of the cylinder.

Hydraulic Oil Supply

What about the supply of the high pressure oil which is required in a hydraulic system? It must come from either of two sources, an accumulator or an electrically driven hydraulic pump.

The accumulator principle is shown in Fig. 4-8. You see it is built somewhat like the spring piston except the plunger has no output shaft. The piston here is sealed tightly against the walls so that oil cannot get out except through the input-exhaust port. This piston is then forced back against the spring when the front of the cylinder is filled with oil. The spring force now is ready to cause the oil to be exhausted from the cylinder under pressure.

The advantage of this system is that oil may be stored in a relatively small "bottle" or cylinder, and this might be placed inside

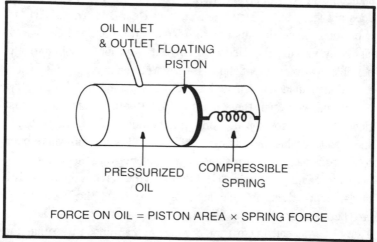

Fig. 4-8. A hydraulic accumulator.

the robot. A disadvantage of the system is that when the oil is to be changed, the accumulator also needs new oil. And you must provide some type of container to catch the oil as it is used, as it does not go back to the accumulator. This is an open-type system.

If you use an electrically driven hydraulic pump the oil will go to a reservoir and from there it can be pumped into the valve and cylinder(s). This is a closed system, which you do not have to recharge. But you do have the added weight and space requirement of the pumping unit.

Now let us examine a modern hydraulic system which uses a transfer valve to control the flow of high pressure oil into and out of a power piston. Such a system is shown in Fig. 4-9. When the valve is moved electrically—and it takes very little power to move it—it will let the high pressure oil flow to one side of the piston and at the same time it will uncover the exhaust port open to the opposite side of the piston. So, as oil flows into one side, it comes out of the opposite side of the power piston. The valve opening can be gradual so a controlled flow is obtained. When the valve is centered no oil flows at all, and the piston is locked.

This is a very fast system. The valve can be operated by small electric currents which energize solenoids on each side of the unit. The spindle itself is normally spring loaded to neutral. The valve can be operated by solid-state devices (transistors), which make a small and compact control arrangement. The valve is normally tight and does not leak, but the oil must be kept clean and free of dirt and impurities. Since the amount the valve can be opened can be adjusted by varying the electrical current, a proportional control is possible over movement and pressure. You can move the piston just a fraction of an inch if you so desire, or several inches, if necessary. Notice that you open the valve to move the piston and then must close the valve to stop motion. But the output from the piston is linear and variable and thus may be the kind of motion wanted in a robot, especially in its arms and fingers. Notice that the fact that the output force may be controlled—that is, if the valve is opened only a tiny amount—then the small input pressure times the larger piston area becomes small, and so, if you use the piston to close fingers, for example, you can make that pressure small enough so that the fingers will not crush an egg.

152

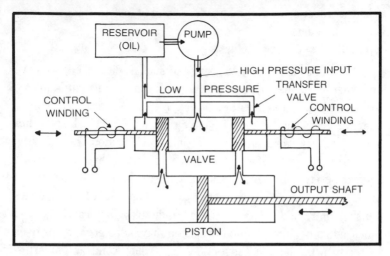

Fig. 4-9. A hydraulic transfer valve and piston.

On the other hand if you open the valve port fully you then might have enough pressure on the piston to make it bend a steel bar easily and surely. With the small opening, there is another factor we should mention. That is a slow movement. This is good if you don't want to grasp an egg quickly. If you do want fast movements then you open the valve fully, and you get increased force as well as very fast piston motion. This latter condition may have good application in tasks where you do want large forces and fast motion from, say, a hand and arm.

Oil Supply

Since we can have pressure multiplication through the use of a small inlet port to a large area piston (relatively speaking of course) we can have a relatively small amount of pressure to operate the piston and still have the required force output. This means we can use a small pump in many cases. Also this means we can use flexible lines for the high pressure oil to convey it from the valve to the piston if this makes for a better arrangement of parts. This might be true in the case of the upper arm, forearm, or fingers of a robot. The flexible line makes it possible for the arm to move in its multitude of directions and still have the strength and quickness and proportional motion that we ultimately desire. In some cases, though, a very large amount of high pressure oil or very high pressures (such as in industrial robot systems) make the use of steel lines necessary to contain the fluid.

Hydraulics vs. Geared Motors

With an adjustable piston system, such as discussed, we have an immense number of fixed positions as the piston shaft is moved from neutral position in either direction to its extremes. When we consider gears and gearing we have to think of backlash and other sources of play which are not possible to eliminate. Thus, for the most precise control of a part of the robot system, hydraulics would seem to be a requirement. In this case the fineness of the adjustment of the moved part is only a function of the sensitivity of the feedback sensing unit, and that can be in millivolts, or fractions of a thousandths of an inch. Such precision may be as great as or even better than the human sensory system which controls our arms, legs, toes, and fingers.

It would also seem, since we do not have gear ratios to worry about, that the hydraulic system might be simpler to instrument in a serious robot effort. But we won't belabor this point because you might find an electric motor system ready to use. In view of its simplicity, with batteries for primary power and switches for control, this might be better suited to your needs. It would not need a hydraulic oil supply with its reservoir and pump. Electricity might be the simplest and cheapest way to go on your own construction project.

PNEUMATICS

The pneumatic system is about the same as the hydraulic system, using similar valves and pistons and the same pressure multiplication principles. It will also require an air-pressure reservoir and a pumping system to fill it. It has been said that the air system is the fastest possible system because the mass of air is less than that of oil and so air can move more quickly. Air is also said to be a cleaner fluid in operation. Ever seen the oil leaks under your car? But one possible disadvantage is that the air is compressible, as we have previously stated. Thus some fluctuation in load position is possible as the size and weight of the load tends to vary.

There is one other consideration of a pneumatic system which is important. The valve tolerances have to be tighter than for hydraulic systems. The same kind of O rings are used to seal pistons and valves but the seal might be better to prevent air leakage. Then, too, the air must be pure and dry. If it isn't the valves may stick and bind and otherwise cause trouble. This means the system must have an air filtering system and drying system. This is more cost. You should try to arrange the system so dust and dirt cannot easily get into it, which means some forethought about its location.

It is possible that in some applications using pneumatics a certain sloppiness may be permitted. The leaking air is made up by furnishing a constant supply under pressure. This may work out okay. We have seen some automated machines in meat counters in grocery stores which use air to seal and stamp prices on packages. We have also seen some applications in a test factory in Japan where the air pressure is used to cause claws to clamp and release objects which the robot arm picks up and moves or assembles. But as we think about the system here, we simply say that before going this route of construction, be aware of the possible increased maintenance costs and construction requirements of the pneumatic system over a well built hydraulic system. Really, the choice of the system you use to power your robot will depend upon the robot's application.

SOLAR CELL POWER

When we consider exotic types of primary power sources, one of the first which comes to mind is the solar cell. These might be placed at particular locations on the robot's body so that they can receive sunlight and keep batteries charged. Or, it may be possible, in some geographic locations where sunlight is always available and intense, to use solar cells to directly power the robot's machinery. It is a thought and it is possible. We all know the satellites use solar cells to power instruments and keep batteries charged.

FUEL CELLS

Some batteries, such as were used by the astronauts, use an organic "fuel" which is chemically handled in such a manner that when brought into contact with plates of a battery type arrangement it causes the plates to produce an electrical output. Algae was one fuel for a battery of this type. Here, all that is needed to have the battery continue to supply electricity is to supply it with "fuel" until something wears out. Maybe this is another possible arrangement for robot power.

NUCLEAR CELLS

The most dreamed of power source is one which is light, small, and rarely needs charging. For the lifetime of the robot, it would always give enough electrical power to operate whatever systems are used. Remember that even with hydraulic operation you need electrical power to pump the hydraulic fluid and keep it pressurized. In a robot you would probably always find a combination of systems, electrohydraulic or electropneumatic for the ultimate level such as the android or cyborg.

The nuclear cell is the answer for primary power. Such power units are already being used in some satellites, where they are far enough away from people so that any problems of atomic leakage and so forth won't cause problems for humanity. The nuclear cells are small and sealed and work for years and years and....

Some nuclear power plants use thermocouples, a combination of metallic elements which, when heated, supply electrical current. Others are small superheated, closed-cycle steam plants which drive electric generators to furnish electrical output. There are other methods of producing electricity through the use of nuclear elements and still other phenomena under development.

But one day, electric cars may possibly be furnished with nuclear batteries. These would last for the lifetime of the car without charge or deterioration and be absolutely safe. We *will* have robots and androids using this power. It is the ultimate as far as current technology knows. Think about it! It is a fun-mental-project to imagine what one could do with such a lifetime source of electrical energy.

CONCLUSIONS

In concluding this chapter on power sources we now begin to feel that we have almost enough basics and background information to begin thinking about actual robot construction possibilities. But this is a little premature. We do have more ideas to think about and examine. After all, aren't ideas the catalyst to action? How about that!

CHAPTER 5
BASICS OF ROBOT SYSTEMS

In this chapter we want to examine some various automatic systems used in robots and androids, as these are the vital key to what they can and cannot do. A knowledge of the automatic systems also makes possible a direct and scientific approach to the design of robot machines themselves.

It seems peculiar to write of robots as machines. But 'tis true, fortunately, or unfortunately. Also, the android, cybert and cyborg are machines, highly sophisticated ones to be sure, but they are machines. Thus they must answer to all the laws laid down concerning the limits and operational capabilities of machines.

I suppose what we are saying is that we must examine some machine concepts, a little theory and some practical considerations to be able to thoroughly understand our friendly machines. We will also now lay the foundation for what will be required of us to keep our particular machines operating satisfactorily.

Thus in this chapter will be stated some important concepts which, we warn you, will be necessary to understand if you want to have the best kind of robot around your house as a product of your own skill, intelligence, prior planning, and creative ability. So don't give up, but push forward with diligence and determination and let us now see what we can learn in the following pages.

TYPES OF SYSTEMS

Any autonomous machine of the type exemplified by a robot must have one or both of the two basic kinds of systems incorporated in its make-up. These are the open-loop system and the closed-loop system.

In the open loop system the input does not depend upon what happens at the output of the system. In a way it is like firing a gun. The trigger-pull and firing of the shell do not depend on whether you hit a target or not. And that was the express purpose of pulling the trigger in the first place. We are now going to say this another way—to be a little sneaky about it. So think the following statement over carefully: The *output* does *not* depend on the *output* in an open-loop system. There is no word error in the way this is written.

A more intimate view of an open-loop system can be seen in your own washing machine in your home laundry room. Here you put in dirty clothes and set the machine to perform certain operations. It washes, rinses, and spin dries. Then it stops. But it never examines the clothes to see if they are clean. It did its assigned tasks in the order you commanded it to do them, then it stopped operating. Its input did not depend on its output.

An example of an open-loop electro-mechanical operation is a switch and fan. If the fan is in another room and you cannot see it, nor feel its breeze, you will never know whether or not the fan started running (unless someone in the house yells at you and says—Well, you know what they say). The fact that you turned on the switch was sufficient to satisfy the open-loop system input. You did your job and did it well. You are not really concerned with whether the fan runs or not, or in another example, whether a light goes on or not. You will think of other examples of open-loop systems which are closely allied with your own experiences.

The closed-loop system is one in which the input does depend on what the output is doing. Think this over for it is very important. The input depends on the output. That means, for example, that if you set the thermostat in your home for a temperature of 70 degrees, the heating (or cooling) system will operate until the input (the thermostat) recognizes that the output (the heat or cold air) has adjusted the room temperature to the desired setting. All automatic heating and cooling systems are examples of closed-loop systems.

158

Just imagine, if you will, what the situation would be like if the heating or cooling were an open-loop system. You'd either be roasted or frozen, and the system would continue to operate, no matter what, until someone turned it off. In this case whoever turned it off in response to a temperature extreme would become a part of the system, making it a closed-loop type. One system of each type is block diagrammed in Fig. 5-1. Notice that the basic difference between them is the feedback loop.

DIFFERENCES IN SYSTEMS

The fundamental difference in the two systems is that the closed-loop system has feedback information from the output. Whatever the output may be perhaps motion, light, or heat, a signal proportional to this is sent to a comparer unit which essentially compares what the output is doing to what it was commanded to do. When the output has done exactly (or to the best of its ability) what was commanded, the difference in input signal and feedback signal will be zero, and the commands to the control amplifier for action will stop.

Notice that the essential concept in a closed-loop system is that the feedback (and it must be negative feedback) always opposes the

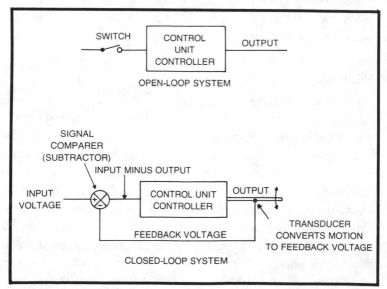

Fig. 5-1. Open- and closed-loop systems.

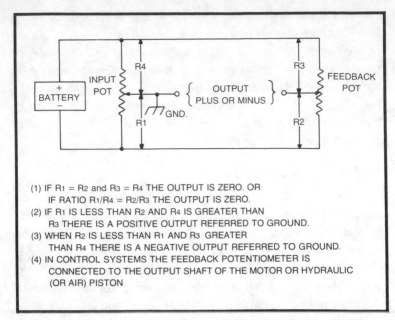

(1) IF R1 = R2 and R3 = R4 THE OUTPUT IS ZERO. OR
    IF RATIO R1/R4 = R2/R3 THE OUTPUT IS ZERO.
(2) IF R1 IS LESS THAN R2 AND R4 IS GREATER THAN
    R3 THERE IS A POSITIVE OUTPUT REFERRED TO GROUND.
(3) WHEN R2 IS LESS THAN R1 AND R3 GREATER
    THAN R4 THERE IS A NEGATIVE OUTPUT REFERRED TO GROUND.
(4) IN CONTROL SYSTEMS THE FEEDBACK POTENTIOMETER IS
    CONNECTED TO THE OUTPUT SHAFT OF THE MOTOR OR HYDRAULIC
    (OR AIR) PISTON

Fig. 5-2. Two potentiometers used to compare input and output signals for feedback.

input command polarity. Figure 5-2 shows how two potentiometers can be arranged to give a control signal to a motor driving amplifier when one potentiometer is the input (a mechanical input) and the second is the output (a mechanical output) transducer.

FEEDBACK

The feedback concept is important. We must consider some other factors of the systems which are vital to the operation of a good, healthy robot. First, the feedback must always be negative. This means it will *always* be of an opposite polarity to the input command signal no matter how the output moves. It *is* possible to connect an output transducer to provide feedback, reinforcing the input command. When this happens the output will drive to its extremes in a violent manner and may permanently damage the machine. This is called positive feedback. It is used in oscillators to make them work.

There are total mechanical types of feedback using gears in nonelectrical machines, but we aren't concerned with that extension of feedback and mechanization here. The electrical feedback signal and electrical input signal constitute the best and simplest way for us to get the required conditions for a good robot design.

If the input is an electrical signal and the output is an electrical signal we can then compare electrical signals in a small, but effective, integrated circuit, the comparison amplifier. This can be done with ease and dependability. But we cannot directly compare a light output or a mechanical motion output to an electrical signal. We cannot compare apples and oranges, so to speak. We must have related items to make a comparison of their size, weight, shape, phase, or magnitude. So we must convert all these other quantities into electrical signals through the use of transducers.

TRANSDUCERS

Because the transducer is required in most automated systems, we had better think about how some types work so that we can have some familiarity with them. The simplest example of a transducer is probably the potentiometer, shown in Fig. 5-3.

It can give a plus or minus output signal at a level from zero (center position) to the maximum of the applied voltage. This ar-

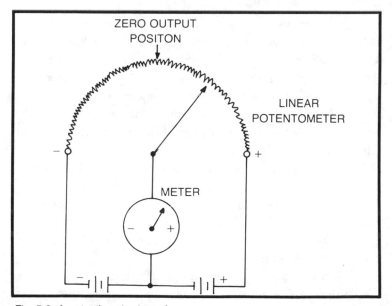

Fig. 5-3. A potentiometer transducer.

rangement can give a voltage which is positive on one side of its neutral (center) position and negative on the other side. Neutral is also the zero-volts output position. In between zero and the maximum we have a proportional (to the amount of rotation) voltage for each position of the potentiometer wiper.

Because each position of the wiper is uniquely identified by a polarity and a magnitude voltage, we could just look at a meter reading and determine exactly how far the shaft of the wiper has been rotated and in what direction, if we know the angular scale in volts per degree. Our control amplifier, in a sense, does exactly this.

It is also possible to use the potentiometer in an ac bridge arrangement. We can have the magnitude of the voltage as before and use the phase (when compared to a reference voltage phase) to tell us the direction of rotation. To obtain one phase at one end of the potentiometer and a 180 degree change at the other end, we feed the potentiometer from a center-tapped transformer. We know that the ends of this configuration are 180 degrees out of phase. The center-tap winding makes it possible to use this in a bridge circuit as shown in Fig. 5-4.

SELSYNS AND SYNCHROS

There are two special rotary units we should know about, the selsyn and, in the automation work, a feedback unit called a synchro. One big difference between them is the physical size of the units. The selsyn is the size of a ¼-horsepower electric motor. The synchro is smaller, about one-inch in diameter by 2 to 3 inches in length. Another difference is that the selsyn actually has power enough to position something attached to its output (receiver) shaft. The synchro is just a small rotary transformer used entirely to give a feedback signal, or in some cases, an input signal. It is just a transducer. Both of these appear in Fig. 5-5. When the rotor of a transmitting selsyn is turned it induces currents through three lines into a receiving selsyn. The receiving rotor is free to turn. It positions itself to exactly the same angular position that the transmitting rotor has been turned to.

When the synchro rotor is physically turned it produces an output signal proportional in magnitude to the amount the synchro shaft has turned. This will be of a phase (referenced to the AC input)

162

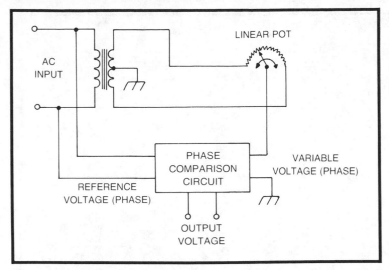

Fig. 5-4. An AC bridge circuit. Phase comparison modules are made by Evans associates, P.O. Box 5055, Berkley, Calif., 94705.

determined by the direction from neutral that it has been turned. In the neutral position the magnetic flux doesn't cut any windings, so the output signal is zero.

THE CONTROL AMPLIFIER (CONTROLLER)

In the output control system for a physical motion the control amplifier always causes something in the system's output to move to produce a signal which will tend to zero the input signal. Since its input is the difference between a command signal and a feedback signal, when the output has done what the input command said for it to do, the difference in signals is zero and everything stops. Actually the feedback signal cancels the input signal at this condition.

An example of a physical motion is the raising of a robot's arm. Assume the input command from wherever it might originate says to raise the arm 10 degrees and that this command is a positive 1 volt. As the motor raises the arm, a feedback potentiometer connected to the arm's joint will begin feeding back a negative voltage to the control amplifier. As the arm raises, this voltage increases until negative 1-volt output from its potentiometer, the arm has moved the required 10 degrees. At that moment the arm should stop moving.

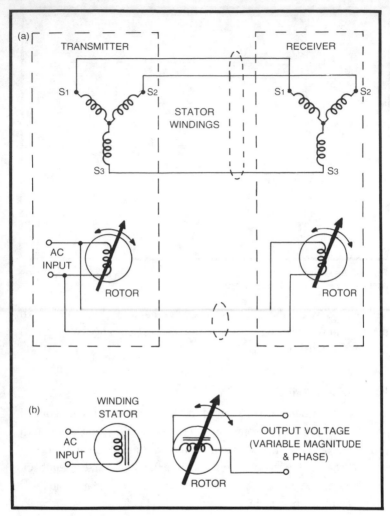

Fig. 5-5. The selsyn (a) and the synchro (b).

Notice in this example that when the command signal originates there is presumed to be *no* feedback, so the motor operates full speed and strongly. But as the arm moves, the difference between the input voltage (command) and the feedback voltage becomes less and less so that if the motor speed is governed by the *amount* of voltage difference, the motor will tend to slow down as the arm reaches its required position. This may be a good thing as hunting may not take place if the slowdown is at the right rate. But also consider that the arm may not reach the specified position, as the motor voltage may be too small to cause the motor to run, even before it reaches a zero voltage value. This can result in a dead zone.

RELAY OFF-ON CONTROLS

You might imagine that in construction of a simple robot, relays might be used to supply power to a motor for an arm, one relay applying the correct polarity for the dc motor to go up and the other for it to go down. Now if the input and feedback signals are equal there is no voltage to operate the relays which in turn run the motor. But when there is a change in input command a relay will be activated to cause full electrical power to be sent to the motor. Now, in this case, the arm moves a full speed until the difference voltage of input and feedback is zero, or low enough for the relay to be de-energized. Note that with the sensitive relays we have nowadays, this can be a very small voltage. Then the relay opens and the arm driving motor stops running.

But what happens in the meantime? The arm has inertia. It was given some speed of movement, so its momentum actually might keep it right on moving for a short distance even though the relay has opened, unless the relay, in de-energizing, has caused a brake to be applied to the motor. This might be either a mechanical or an electrical brake.

Suppose we didn't have this braking. In moving past its required position, the arm would generate a feedback voltage of opposite sign and of a magnitude larger than the input command signal. The other relay would then energize, causing the arm to reverse direction. Again full motor power would be applied to the arm, moving it very fast so it would overshoot the desired position in the opposite direction. Thus there is hunting, or oscillating, of the arm around its desired 10 degree up position. It would be much better to have a gradual and precise movement of the arm to the required position. Finally, on this subject of hunting, or oscillating of the output, can you imagine how a robot would walk, given this kind of leg positioning machinery.

OTHER TRANSDUCERS

We mentioned earlier in this text that pressure feedback signals are needed to adjust a hand grip. We mentioned various types and discussed how they can be used. Now we will consider the signals

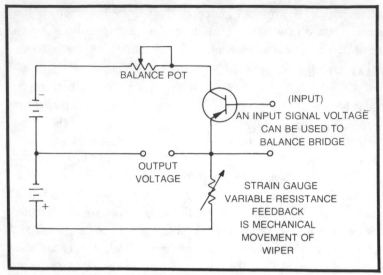

Fig. 5-6. A typical strain-gauge feedback circuit.

out of these elements as feedback signals which will be compared to a definite input command. This will, in turn, originate in the robot or from an externally created signal which you send the robot. The piezoelectric transducer will give a signal when bent or twisted. Because of the circuit complexity we won't elaborate on it here. We will examine its use in some later pages, so it won't be neglected.

The strain gauge or carbon resistor is more to our immediate concern. It can be an element in a balanced bridge as shown in Fig. 5-6.

Here, we are using a transistor for the other element of the balancing part of the bridge. Notice how the two elements, the strain gauge (or variable carbon resistor) and the transistor, take the place of the two halves of the potentiometer shown in Fig. 5-3. If a command to "close fingers" is sent to the robot, or generated within it, the signal could operate the transistor of Fig. 5-6, change its resistance, and cause the bridge to unbalance. This could create a signal at the output with a polarity according to whether the command causes the transistor to increase or decrease its resistance. Now assume that the fingers close and that they continue to close until the strain gauge is distorted enough so that the resistance presented by it also either increases or decreases, according to how it was physically strained. In the process the voltage across it

166

changes until this new value of resistance equals the resistance of the transistor. The bridge is again balanced, the output is zero, the hand stops closing, and does now exert any additonal pressure on whatever it is grasping. We have accomplished one kind of force feedback. Please realize that this force feedback may be just a part of the control system really causing the closure of the hand; other types of feedback may also be present.

Of course when the internal or external command to the transistor changes, perhaps to command the hand to release the object, the resistance of the transistor again changes. The hands motor, then receiving an opposite polarity signal from the output of the bridge, must release pressure, causing the strain gauge to relax back to its original shape, which will then rebalance the bridge and stop the hand's grasping action. The movement of the wrist and hand is an additional area of control as we will find out in a later chapter when we discuss construction.

PROGRAMMED SYSTEMS

When considering programmed robots we think about those preprogrammed systems built into them which, when activated, will cause the robots to do certain things. We often hear the word programming. It is used in science fiction stories about robots, and we also hear it when we discuss the new micro- and mini-computers now available on the market.

Probably the most basic internal preprogramming system is a timer-actived set of switches. These might have, very fundamentally, the appearance of Fig. 5-7. While this unit is mechanical, it is operable and reliable and will do the job. Of course we could get the same action a little more efficiently and with less space and more operations if we used a timer operated by digital circuits. Let's examine how such a timer might be used.

Imagine a preprogrammed action such as having the robot speak when it hears a sound. The sound will activate one relay switch. This relay might start a motor driven cylinder, as shown in Fig. 5-7, to begin rotation. The relay starts the motion. This is continued by a contact on the rotating drum. After one revolution the circuit is broken and the relay must be energized again to repeat the action.

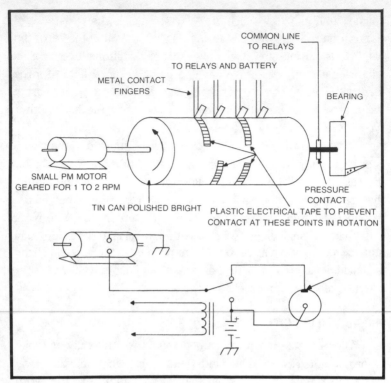

Fig. 5-7. A basic timer.

Attached to the cylinder by pressure contacts are a set of fingers which can, in the simplest case, merely be some stiff wires making contact with the tin can drum surface as shown. When the tape insulation is under the fingers there is no contact. Or you might use instead a set of plastic cam lobes on the drum, which close microswitches switches when the lobes move under the levers.

In any event, as each switch is closed, it can cause other electrical parts to function, which in our case of the voice response, would be a recorder. This would then give a prerecorded message. Note that the drum switch has to make contact for the length of time the voice is heard otherwise the recorder will be turned off.

Now at the same time, the robot can be programmed through the other switches to move and raise its arms, giving the appearance of real live action. Its eyes could flash and the head turn. And all this would happen after you speak to it and say, for example, "Do

Something," a two-word code. Think of the fun you could have by having the proper recorded message spoken by it. This would be preplanned for the particular group you want to have around when you demonstrate your robot. Notice that the primary requirement here, aside from having the multiswitch control operating, is that the voice control unit must start the timing motor turning the drum (see circuit of Fig. 5-7). Then the timing motor must continue to operate throughout at least one revolution of the cylinder, operating the control switches.

We grant that this is not a thinking robot, but it can do many things to the amazement of your friends and relatives. And it would be very simple to build, as far as the primary action system is concerned. It would also be possible to adjust the sound energizing system so that if you spoke "Halt" it would stop all action. Then saying "Now be nice" would cause a second program timer of the same physical construction to operate, causing a new line of patter (in the same voice of course) and new actions. So with a little practice you begin to have a robot which does seem to respond to your commands and carry on a conversation with you.

To make the robot seem to talk, you prerecord the program you want, leaving pauses which you then know are there. During those times you can speak your own sentences which fit into the program, making it seem like you are answering the robot or it is answering you. It takes a little doing, but you can practice it, and it's a lot of fun. You might write a script for the robot and record it with pauses at the proper times and places you would speak. For example,

You say   "Do something."

Robot   "I am doing something. What do you want
               me to do?" It flashes its eyes and
               waves its arms.

You   "I want you to serve my guests a drink."

Robot   "Don't be foolish. I am not programmed to
               serve just anyone anything."

You	"I'll change your program."
Robot	"Are you threatening me?" Eyes flash brightly and it moves forward for a few seconds.
You	"I've changed my mind."
Robot	"Good thing you did too." Arms stop waving, eyes dim, and it stops moving.
	(Get someone else's voice on the recording. Don't forget that.)

You would time your program recording so the pauses would be long enough for you to speak, and time the actions to resume when the robot recording mechanism begins to operate. Get the idea?

Now just for identification let's call this system a preset system because all actions are preset into the robot, and they will take place in a sequence inside the machine.

Let's consider an expansion to this basic idea. This would have the voice system so adjusted and connected that it responds to the number of separate and distinct words you speak, perhaps up to four. If you speak continuously the circuit will do nothing. But if you pause for a definite time and then speak slowly and clearly with good word separation either one, two, three, or four words, a decoder and timer can be arranged so that you obtain different robot responses for each code-word group. It's fun to think how this might be accomplished. One final thought which might be useful here is that if you have trouble with a coded-word, voice-activated system you might use a concealed high frequency sound generator or wireless transmitter which won't be heard and can't be seen by your guests. With this unit you could send signals to activate the programs in the robot in accord with some code. This way you can say anything you want to say for any length of time and then activate the robot's response.

There are probably other switching systems you can think about which operate in this manner to activate various responses from the robot. You should investigate these. Perhaps you'll find some flip-flops, solid-state relays, or solid-state timers which will be small and compact and can work out just as well as the mechanical timer we've shown here. But our little timer is inexpensive and easy to construct. You might start by using it.

INTERNAL-EXTERNAL SYSTEMS

The preprogrammed robot just described is an example of an internal program which is set into motion by a command from its master. It is also possible to have an internal program set into operation by some other kind of external stimuli which the robot can recognize through its sensors. One we immediately think of is a light sensitive photocell which no doubt you are already familiar with.

Now you might have a nice, well-behaved robot in the house standing guard at night by patrolling a given path. It could be equipped with an accoustical sensing system which would sound a loud alarm through an internal speaker in the robot when activated. This robot might use a daylight-dark sensor to start itself into action whether you are there or not. By using a definite preprogrammed path the robot could move itself through various rooms or circle a given room continuously until daylight caused it to stop its internal programming action. This might be the simplest watchman for your home. To avoid having a flashlight stop it (burglers do use them) you adjust its timer so that once the system is activated by nightfall, it continues operation for, say, eight hours before returning control to the light sensor.

Some thought might produce a more elaborate system in which a path, such as discussed in an earlier chapter, is laid out to be followed when the robot is programmed to start moving. A better system would not get off track in operation. In the first case we mentioned here, the robot would have to move forward for a given time or distance, turn, move forward again for a given time, and so forth. Now, if its speed varied at all, or somehow (and it always seems to happen) the robot got slightly off time or track, then it might begin to bang into things and the Empress of your household would take a very dim view of your experimentation. You might find yourself sleeping with the robot instead of her.

So be careful, be sure, and don't mess up the Empress' furniture or household by operating a dumb robot that doesn't know where it is going or what it is supposed to be doing. Perhaps the track-following method with the required sensors to stay on a safe pathway would be best. The photocell acting on the reflected light from a white strip could be one of the simplest ways to instrument this idea. But this will also require some experimentation to perfect it for your location. Perhaps you will come up with a better method. We take our hats off to you for that kind of thinking.

RECORDING PERSONALITY

We hinted earlier that your robot can have a personality. It is not only possible, it is probably going to be a requirement. You might begin thinking about this phase of your robot's development by studying people you know who have a decided personality in their voice or speech pattern. Close your eyes and listen to them talk and you'll likely find the voice (male or female) most pleasant to your hearing. This is the personality you want the robot to have. Then you can have that person read the dialogue your robot is to speak.

It is interesting to think about how you might record the robot's speech. If you use a cassette it might be hard to get into the middle of the tape to playback a portion of that middle segment. It would be more efficient to use one of the small computers now in use with a floppy disc recording system, or perhaps a quick-access tape. A micro-floppy disc system has a head which can be placed down on the recording at any point with an exactness of ten-thousands of an inch. It makes this placement quickly and precisely. It can instantly pick out any statement or response from anywhere on the disc.

By the way, on the market as of this writing is a toy telephone that responds when 1 of 10 buttons is pushed. Some elementary mechanical system must be used to position an arm for playback. That would be less costly and might bear looking into. Relays give physical motion, and can be electrically operated. So they might be somehow used to position the reading head of the system.

When we think of the thousands of responses which might be recorded inside the robot in some manner, not excluding a computer memory system, we can begin to understand how a robot might have a very large vocabulary. Did you know that the average person's speaking vocabulary is only around three to five thousand words?

We become excited when thinking about the possibilities of an advanced android able to select appropriate words from a memory in response to something said. It would select the words and sound them in proper sequence to make up sentences. This differs from just recording statements and questions. The electronics inside the robot would have to key-in on your words, analyze them, and then find words in its memory to make up a suitable response. Some experimental work doing this actually has been accomplished at some universities. But for us, the fast, accurate positioning of a playback head or tape readout combined with a minicomputer makes experimentation along this line possible, but expensive.

For your first robot consider a small, more economical tape recorder and playback system, or just a playback unit. You use the recorder at home to make tapes which can be replaced easily as the need arises.

SPEAKING BY REMOTE CONTROL WIRELESS

Of course you can do this. You must have a transmitter attached to a concealed microphone on your robot. The sounds can be transmitted to a remote station where you have a receiver. A transmitter, there, on a different frequency, sends your voice back to a receiver in the robot. It talks through a concealed speaker. You can then have your robot carry on a conversation with anyone it may encounter when a distance away from you. It will seem as though the robot is acting entirely on its own *if* you cannot be seen during this process.

The transmitters can be of low power because you will be within 100 yards of your robot in most cases. You might also control the robot's actions by radio control, using another transmitter and coded signals for the means of contact. But that isn't an autonomous robot. It is fun, but it isn't like having the robot actually do things on its own with simply a stimulation by voice, position, or the action of something near it which sets it into motion and operation.

MULTIPLE SYSTEMS

We begin to see that inside our robot we may have many separate and distinct operating systems—such as the drive, the voice, the arm movements, and the steering and path following systems. Each one may be independent of the others, or we may have many which are dependent upon what another system is doing.

With interdependent system operation we are approaching the time when we might consider giving our poor old robot a brain so that it can keep track of everything and time things properly. He might need a memory to remember what you told it to do, when to do it, how to do it, and to find out when it has done it. Of course we must remember all these things also.

We will be needing a small computer. But don't throw in the sponge yet. We will tackle this problem with dexterity and try to make it as simple as possible, perhaps too simple for some. And if so we ask forgiveness. And we will proceed with caution as we have tried to do on everything else in this work. We will try to solve all the computer requirements and operations easily and directly—you, the robot, and I.

CHAPTER 6
THE ROBOT "BRAIN"

As we saw in Chapter 5 a robot can be programmed with a very elementary cylindrical drum. This is so constructed that it is part of the switch or it can be made to operate microswitches. No doubt you have already thought about this and decided that for a really intelligent robot—one that would meet the approval of the King Robot himself—we must have a much faster programming system, one that will permit the interconnection of many more circuits in almost any desired sequence and for any desired times. We would call this internal control unit a robot brain. Ideally it would be a microelectronics circuit card which, to most of us, would seem capable of controlling all of the robots many electrical, electronic, and mechanical sub-systems.

To understand just what it must be able to do and how it works, we must now do some serious thinking about how a computer works, especially one able to handle many inputs and do almost anything with the signals representing these inputs. It must interface with whatever machinery might be attached to the output. Keep in mind, however, that the little programmer may be used as a guide to our thinking.

Let's take a moment right here to examine a concept of a recognition system for class II robots as defined by Klingman the Third of the Marshall Space Flight Center. We will also be exposed

175

to his thinking and technolgy which defines robots in terms of class I or class II types. He said:

Class I robots are designed to function in an unknown environment. Class II robots, on the other hand, are intended to function in a specific environment. Examine Fig. 6-1.

Class II robots thus entail the concept of organism or environment design rather than just organism oriented design. Because of this, it is possible to provide, through appropriate planning of the environment, a robot with the ability to recognize selected environmental features. This can be achieved by associating with such features as infrared, visible, or or ultraviolet sources which emit code identification signals.

Numerous means of coding exist, among which is frequency, or pulse-rate, encoding. The receiver is located on the robot and may incorporate a means for determining the direction of the signal source. Several means exist, including both active and passive direction finding techniques. They are based on an extended detector located behind a slit, or aperture. The output of the receiver is amplified, if necessary, and the frequency, or pulse rate, is extracted either by counting or by active filtering.

Both digital and analog implementation are feasible; however, the digital system does offer more flexibility. The binary output of a counter is compared with stored binary words, and an indication is provided if equivalence is established. The counter is gated on and reset with timing pulses provided by a system clock. The information in storage need not consist of fixed words corresponding to known objects and locations, but may consist of a recognition algorithm (formula) which combines the robot's current state with its past history to generate a recognition against which decoded incoming signals can be matched.

This system is superior, for class II robots, to the three major techniques of robot control used by prior art. It has inherently greater resolution than do systems based on radio or radar signals. Its data storage and processing requirements are small compared to those systems which are dependent on totally self-contained methods of position-data acquisition via comparison with a computer stored map (the map matching technique we discussed earlier). And

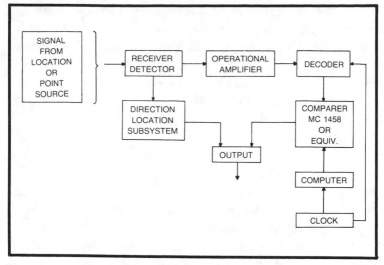

Fig. 6-1. The basic system of a class II robot.

finally, it produces results presently unobtainable by systems that attempt to implement various pattern recognition methods.

Thus spoke a wizard. His philosophy is to have the robot recognize some phenomena of the terrain and identify with it. And it would be passive in concept as opposed to the active system which sends out pulses and acts on the reflections. The passive system acts on signals received from some other source rather than those it produces.

Well, we've added some more information to our minds about robots and now realize that there are such things as class I and class II robots. Kryton is gripping my shoulder and I'm certain it's to make sure I don't write him down as a class II type.

Once again, why must we consider the more elaborate computer in our robot? The answer is that we may want to have various inputs into the machine and will want to have all kinds of responses—physical, electronic, and whatever—to those stimulation signals. We want the ability to be able to examine the input from several sources, and depending upon what each source (sensor) tells us, decide what action to take or how to respond to the situation.

Let's imagine some examples. Suppose we programmed our robot to take the dog for a walk. He can do this nicely by going down the sidewalk with the leash in his hand—to the utter consternation of all of our neighbors—to the end of the walk and then turning around and retracing his track (steps?) back home. Now suppose that while doing this it suddenly begins to rain. A new element has been added. Under this situation we might want our robot to turn around quickly, perhaps right then, and hurry to get home before he got soaked. The dog might yank harder at the leash, and so the robot would want to tighten his grip on the handle. So we would have several new conditions governing what the robot must do.

Or take another example. Suppose the robot is walking the dog and someone stops to ask it a question, just out of curiosity to see what will happen. It would be nice to have the robot stop—perhaps pulling in on the leash so the dog won't bite that curious bystander—and then respond through a sound activated relay system with a recorder message. He might also want to turn toward the bystander. This could be done with sound sensors (or sensors using some other phenomena to give directivity toward the speaker) to pinpoint the direction of the voice. The robot then would respond with a suitable answer.

"Well," you say, "What if a truck passes or there is a sound from an airplane or some other source. Suppose the dog barks?" Granted. Tis true that might be a problem unless some computer element inside the robot is so designed that it accepts information from an outside source only if it resembles a human voice. That might be done using electronic filters. Thus it should not respond to other noises. Possible? Yes. Difficult? Perhaps. Desiging a circuit to do this can and has been done.

In this light you can perhaps imagine many of the things a robot brain must analyze and act upon to give reasonable responses to your new household pet. This requires a very fast acting computer to put all the sensor and preprogramming information together in comparison circuits, rejecting some because of low voltage levels, perhaps, acting on others because of their higher voltage levels. It would determine priorities based on the size and maybe the shape of the signal inputs.

If you want a really mind-challenging operation imagine the coordination required and the problems involved in providing independent movements of the robot's arms and hands. With humans the movement of one hand, arm, or finger (or all of these) depends to a great extent (or entirely) on what the other extremities are doing at that second.

When you reach for a pencil on the desk, your eyes identify where your hand and the pencil are and then cause movement of the muscles (try to imagine this movement in slow motion) to include opening the fingers at the right instant to grasp the pencil, adjusting it in the grip, then tightening on it to pick it up. Try to pick up a pencil with your eyes closed. See what happens.

Yes, we will need to know a little about computers, generally how they perform their operations using processors, instruction units, clocks, memories, and the like. Bear with us and we will try to unravel the mysteries, making everything as simple as possible.

There is a word of particular importance to those about to make use of a computer. That word is interface. It simply means a connection between the computer and some other device, and goodness knows we will have lots of devices to interface with the computer. When speaking of interface we simply mean circuitry able to match the computer with other equipment. You may want to expand your knowledge of computers with a study of the many texts now available.

THE BASIC COMPUTER PROCESS

Let's start by taking a good look at the basic computer process in Fig. 6-2. This shows what a good computer designed for robot use might look like in a very simple block diagram form. As we can see there must be some provision for inputs from the various sensors—both primary and feedback types. There must be an input for program instructions the robot must accomplish. There must be a timing system so all these signals won't get confused and can be examined one at a time—and in a very short time at that. There must also be a storage (memory) system to hold the signals until they can be examined, used, or erased.

There must also be an output for the various signals controlling the output devices. We must have interfaces between the input sensors and the memory storage/computation unit, between the program unit and memory, and between outputs and memory. These might be cables designed to connect between terminals of each unit in such a way that impedances on both ends of the cables are matched. Or we might have special matching circuits.

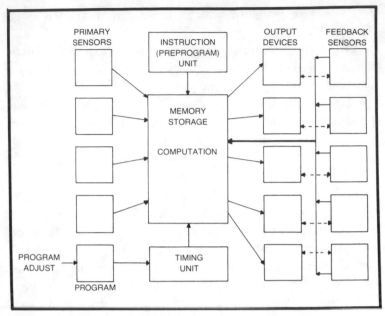

Fig. 6-2. A basic computer for a robot.

BREAKING IT DOWN

Let's go even more fundamental now and examine a single sensor and a program recording feeding an output. The diagram appears in Fig. 6-3. Before we examine this diagram, let's look at some symbology used in computer circuits, just in case you aren't familiar with it.

BLOCK DIAGRAM SYMBOLOGY

Block diagram symbology, shown in Fig. 6-4, consists of the basic building-block gates used in digital computer applications. For the moment they are all that we will need to examine the fundamental operation of our robot systems. A gate is a circuit which will pass a signal only under certain conditions. It does not supply a continuous signal like an amplifier or power supply. Always look for the conditions under which the gate will and will not operate.

THE BASIC VOICE CIRCUIT FOR A ROBOT

In Fig. 6-3(a) we see an input of sound (voice?) through a microphone—although one could also use radio, light, heat, or infrared, with appropriate sensors—to an amplifier and relay. When the relay closes it causes a voltage to appear at its output. This voltage, a high state, is presented to one input of AND gate U1A. At the same instant the high state goes to AND gate U1B and to a clock circuit. Remember that a clock circuit is essentially a multivibrator, or a series of them, which will produce pulses at regular intervals, depending upon how its time constants are adjusted. We will assume the clock pulses are long enough for the programmed recording output to make a one-sentence replay in response to the input sound.

The clock output, a voltage, supplies the other input for AND U1B. This is the second input for AND U1A. This drives the output of U1A high, causing the recording to play and continue playing as long as all voltages (high states) are present to both AND circuits. Notice the difficulty with this arrangement. If the input sound ceases, there will be no output as the relay will de-energize. What do we do about that?

Aha! You guessed or analyzed the solution correctly. We must either have a self-locking relay on the voice amplifier output, or we must provide a memory to keep the voltage present for a preset time. We might even arrange the circuits so they are clock controlled as shown in part (b) of the figure. In this latter case the clock will do two things; it will de-energize the voice relay sensor section while the output is playing, and it will make the output recording continue to play until a specified time has passed. This prevents any other voice sounds from causing an error in the sequence. Once the recording has played as long as necessary due to clock timing, the clock holding pulse vanishes. The whole system is made ready for another input as the clock re-energizes the input section.

"But," you might say, "That clock isn't really a clock. It is a simple time delay circuit." You are quite correct. A time delay circuit, part (c) of the figure, could serve the purpose just as well in this case. We have indicated a clock because it does do timing, and that is what the clock circuit in any computer does, no matter what circuit it may be.

Be sure to consider a timing circuit. A simple reset multivibrator provides the basic circuit for this operation. Voice operated relay kits (Radio Shack) are easy to build. The AND gates are available at parts stores. You can build this circuit for use around the house perhaps to answer the door bell, a knock on the door, or something like that. In a robot it could be used to answer people who speak—not always kindly—to our humanized machines.

Part (c) Fig. 6-3 shows the simplest possible method of obtaining a response to a word or tone input. Here we might have a delay multivibrator or a high resistance relay coil with a large capacitor across the winding. Either of these allow the recorder to be activated and run as long as the delay circuit holds. Of course, in the delay circuit there should be a disable contact or another circuit which will disable the input sensor while the recorder is running. It's not difficult to expand on these basic ideas. So why don't you have a go at it. You'll probably come up with some dandy circuits of your own, not only for voice but also for other inputs as well.

FURTHER EXAMINATION OF ROBOT COMPUTER REQUIREMENTS

Basically what we desire in a robot computer is lots of inputs and outputs, and the ability to have prearranged programs perform various tasks. As a starting point we might make a list of these tasks and then break down each task into requirements for motion or other operations required to accomplish the task. Generally we will find that the number of signals and movements (or operations required to perform even simple tasks will be far greater than anticipated.

With this in mind let us list some tasks we might want to have our robot perform.

1. Become energized when spoken to.
2. Move in a preprogrammed direction when ordered.
3. Turn head to face source of sound.
4. Flash eyes while speaking.
5. Move arm out to point.
6. Return arm to side.
7. Raise arm in response to command.
8. Stop movement upon sensor recognition of position.
9. Speak in response to specified questions.
10. Raise both arms and lower them to sides.

As with all the information in this work, we hope that this short list will titillate you into expanding and elaborating on tasks that may be unique to your own desires and situation.

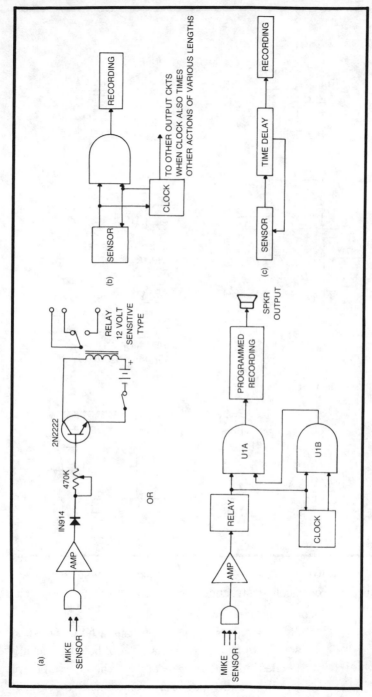

Fig. 6-3. Basic circuits under control of a clock.

183

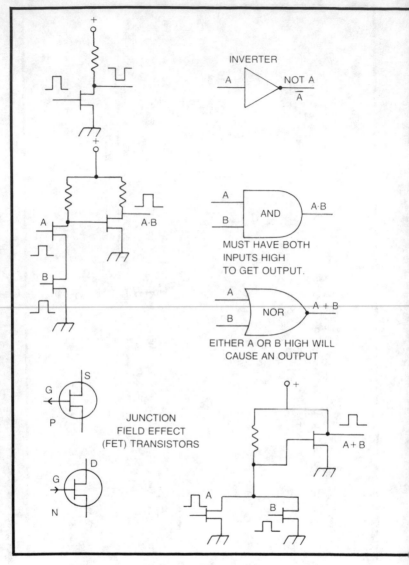

INVERTER

AND

MUST HAVE BOTH
INPUTS HIGH
TO GET OUTPUT.

NOR

EITHER A OR B HIGH WILL
CAUSE AN OUTPUT

JUNCTION
FIELD EFFECT
(FET) TRANSISTORS

Now to continue our example, we expand on these tasks to see what control signals (instructions) and operations will be required.

1. Become energized when spoken to.
   a. Requires a voice sensor, perhaps a filter too, so all sounds won't energize it. A delay or holding circuit is needed so that once energized the operation will continue until specifically commanded to stop by

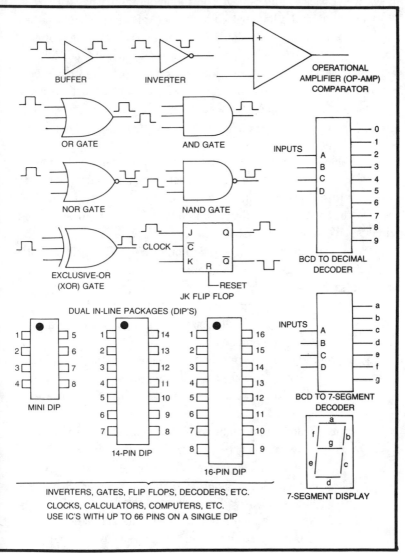

Fig. 6-4. Some basic gates used in computers.

voice or sensor signal. Must set all moving parts to neutral at moment of energization and clear all previous information except permanent programs from the operating circuits.

b. Must signal that the unit is ready to function.

Let's begin with this situation and imagine what controls we will need. Refer back to Fig. 6-3(a). The sound operated relay will have to operate several additional circuits, one of which should be a clearing, or neutral, signal along a common bus. We must provide that there is a difference between a de-energized, or off, condition, and just a pause between operating times when the robot might be still but is energized. We don't want to wipe out all instructions just because the robot paused for a moment. Thus the circuits must discern the transition from inactive to active from that of delay or pause.

Considering all these, then, we assume three basic things must take place. First, the voice circuit must recognize only a specific sound or tone. Second, it must clear everything not permanently a part of the system. And third, it must apply power to all other circuits. We must assume that power is always applied to the on-sensor input so that it *can* activate other circuits. Finally a light must flash, eyes light up, or a spoken response come out of the machine saying "I'm ready. Are you?"

2. Move in a preprogrammed direction when ordered.

    a. Requires a de-energization of input so commands won't be confused.

    b. Requires energization of sensors locating path.

    c. Requires turning and orienting itself on path.

    d. Requires drive wheels to move at specified or preset speed.

    e. Follows path until sensor identification with something causes it to stop or turn and come back along path to starting point and stop. It might stop at various places along path if a timing or distance measuring wheel is used, and this is desired.

    f. Neutralization of all motion circuits and energization of primary circuit so it will be sensitive to another command.

3. Turn head to face source of sound.

    a. Requires a comparison circuit (audio, radio, light, infrared) to be able to "direction find" the source of the sound. Separated sensors required like ears on our head.

    b. Requires motion circuit with feedback to turn head until sound balances out, or nulls.

4. Flash eyes while speaking.
    a. Requires a connection in audio responding circuit to energize lights (eyes) in head which will operate with variable intensity in proportion to sound being emitted. Will stop when sound ceases.
    b. Requires neutralization and disconnect when speaking is accomplished.
5. Move arms to point.
    a. Interconnection (if desired) with voice or motion circuits so that arms raise to 45 degree level at the moment of energization. Will hold there until a short time elapses.
    b. Arms must return to sides under internal program after short time.
6. Included in requirement 5.
7. Raise arms in response to command.
    a. Requires a separate identification (decoding) circuit which will recognize just the command "Raise arms." Motions of the robot might be then a one-word code or a three-word code.
    b. Requires de-energization of all other circuits, (except perhaps lights (eyes) flashing when raising arms. Must be able to lower arms on repeat of command code such as "Drop arms."
8. Stop movement upon sensor recognition of position.
    a. Might be used when telling robot to back up a couple of feet or go forward a couple of feet. It would not necessarily be following a path.
    b. Might want robot to turn in some direction without much forward movement.
    c. Requires a rotating wheel (like odometer in car but sensitive to distance measured) and command which must be decoded internally into proper distance.
    d. Requires a means to measure rotation of body of robot with respect to a fixed direction. Use of gyroscope reference may be required.

9. Speak in response to specified questions
    a. Here, some coded words might be used which have a particular sound content to energize certain segments of prerecorded speech from tape or disc record. With enough speed and proper programming the robot might select just words from various places on the tape or disc and speak them through its loud speaker. Must be fast enough to make an intelligent response. A computer may be necessary.
    b. Requires decoder, internal programmer, and very fast action if words selected. It might be easier to select recorded sentences.
    c. Must put all other circuits except eyes and perhaps arm movement circuits in the off or standby condition.
10. Raise both arms and lower them to sides.
    a. Needs a decoding of command and activation of both arms and eyes. Might speak when doing this.
    b. Requires reset of all circuits to start or neutral position when action is done.

SO WHAT?

Well, the so what of all this is that as you make up your list and begin to itemize each requirement, you begin to lay out the foundation for the equipment you'll need and the type which must be used. There may be dual-purpose circuitry which can simplify the actual construction. But basically you must know what you want to do, how you want the machine to operate, and in a general sense what kinds of circuits and devices can enable you to accomplish what you want done. This must be known before you can even begin to construct a robot. Of course you might allow room for growth of the system, adding things which you did not think of the first time around or which you come to see as possibilities after some parts work as they should. It's really a challenge and lots of fun to work out a paper study of a robot system. Then finding off-the-shelf items and constructing the mechanisms and circuits is the fun part of the job.

THE SCANNER

It isn't out of place here to mention the scanner system used with CB rigs. A scanner checks all the frequencies, stopping on active channels or priority channels in use. When one of these signals goes away it scans on for the next one. You might consider using such a device to check the various sensors or codes you use in operating your robot. Or you might think of other ways to use such a system in your man-like machine. Sampling sensor outputs could cause the robot to start into action in some manner, depending on what sensor sends out the signal. This might be good for a sentry robot. After all a scanner is just an electronic switching device much like the old mechanical rotary switch which has lots of contacts and a common output. The electronic scanner is faster but looks at lots of inputs and sends them out on a common line. Perhaps you would want to have the sensor output coded. When the scanner receives a signal it will not just send it to its output but will energize some particular circuit associated with that sensor.

SOME BASICS OF CIRCUIT SELECTION

Well, we've come quite a way in this thinking about the computer brain and now it would be advisable for us to consider some circuits which might be used in fabricating such a brain. Let's examine Fig. 6-5. In this figure we show at (a) a simple relay with a normally closed contact, a contact which is closed with signal, and a coil which may have various operating voltage requirements. You select the coil operating voltage for the circuit you have in mind, perhaps a 6-volt coil or a 12-volt coil.

This relay, a basic off-on type, is important because through its contacts you can control the flow of large amounts of power to motors or other things. Yet the relay takes only a small voltage and current itself for its operation. Always ask for the relay contacts current handling capability to be equal to, or larger than, the actual current you want to control. Remember that most equipment draws a large start-up current. Notice that as long as current is applied to the coil the relay contact will be closed and current flows through the n.o. contact. When the relay is not energized there is no current flow to the motor or whatever is connected to this contact.

We see an expansion of this relay idea in (b), where we have two armatures and four contacts. Two of these are normally closed and two will be closed with signal to the relay. This gives more flexibility to control relay circuitry. Of course it is possible to obtain relays with a larger number of contacts and armatures which will be closed and opened like this if your particular application should require it. Just get a relay catalog or ask about them at the radio store.

One of the most basic uses of the double-pole double-throw relay of (b) is to make it into a self-latching switch. If it is wired as shown at (c) a momentary grounding of coil 1 at X will cause the relay to close and it will stay closed until released by a pulse to coil 2, the release relay. With the set of contacts at Y an external circuit can be energized such as a motor circuit and you can have the motor run as long as the latch is maintained.

One limitation to this circuit is that you must continuously supply power to the latching relay to keep it closed. That can drain a battery supply. But an advantage to this circuit is that whenever you supply a ground to coil 1 you get a latch, and whenever you apply a voltage pulse (or ground if it is so connected) to coil 2 you get a release. You can never get these two operations mixed up or out of sync.

Because of the limitation of the electrical latch and its current consumption, we look for an equivalent circuit not requiring constant power. At (d) we see how this can be done using a stepping relay. This, in a sense, is a mechanical latching relay (and there are such things made a little differently). In the case shown a pulse will cause the relay armature to pull down. This pulls on a ratchet wheel which steps the wheel around a precise part of a rotation each pulse. Thus, with a suitably shaped cam attached to the wheel axle, one or several contact switches can be made to close and open at various degrees of rotation of the wheel. You might write for Guardian Relay Co's catalog showing stepping and mechanical latching relays, as well as normal relays. There are other manufacturers who also make such relays.

Since relays such as we have mentioned will be operated by solid-state circuitry, we need to look at a typical circuit. It is shown at (e). This is a popular relay amplifier. With a proper selection of Q2 and Q3 almost any size relay might be operated. Also, using high-current transistors in a similar circuit you might find that to run small PM motors you would not need to use relays at all. Finally we need to mention that there are also solid-state relay circuits which do not have any moving parts. These are usually available off the shelf.

Since we have mentioned a voice operated relay many times examine (f), where one example of such a unit is shown. It is built around the basic 555 timer IC and is adjusted so that the output relay holds closed during normal pauses in human speech and opens when a specific time passes after the speech stops. If we wanted to make a decoder out of this circuit, we could adjust it so that each word would make the relay close. It would open on the slightest possible pause. Thus each word would produce a pulse output and a series of these could be counted and decoded as commands to the robot. You remember how we mentioned that one pulse could mean go forward, two pulses could mean stop, three pulses could mean go backward and four pulses could again mean stop. Think of the one-, two-, three-, and four-word commands which you might say to cause the robot to do these actions and you've got the idea for this kind of control. Of course you'd speak the commands sharply with just a little pause between each word. The relay output might, in its simplest sense, operate a stepping relay motor control circuit.

It is interesting that we've seen a small toy which uses a sound system for operation. It's a small car which goes forward or backward on command. The activating device was a small cricket snapper giving a sharp audible click when pressed. The amplifier in the toy was made responsive to this click and to no other sound. The car could be caused to go forward or backward with alternate clicks. It had no stop. The amplifier here was probably a very high frequency type so as to discriminate between this sound and other sounds in the room. It was not very sensitive and the cricket had to be held close to the car to activate it. This also avoided noise frequencies which might interfere with its operation.

A MECHANICAL ROBOT BRAIN

We can use mechanical relays to fashion a brain for use in a robot. Some people like to see the wiring and action of electromechanical devices. For them, this system is particularly attractive. Other people like to have much smaller and more sophisticated

191

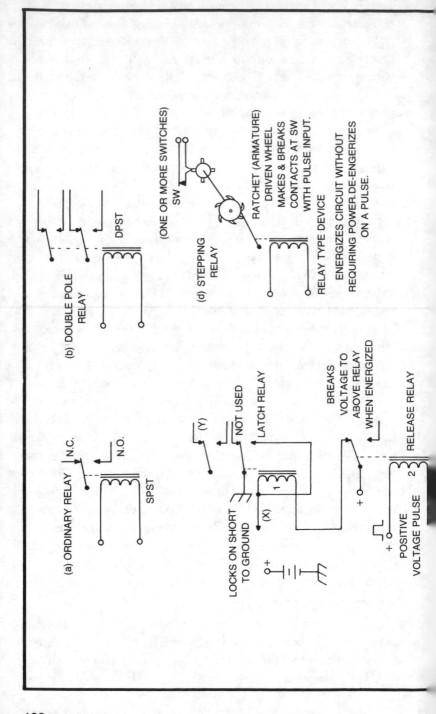

(a) ORDINARY RELAY

N.C.

N.O.

SPST

(b) DOUBLE POLE RELAY

DPST

(d) STEPPING RELAY

(ONE OR MORE SWITCHES)

SW

RATCHET (ARMATURE) DRIVEN WHEEL MAKES & BREAKS CONTACTS AT SW WITH PULSE INPUT.

RELAY TYPE DEVICE ENERGIZES CIRCUIT WITHOUT REQUIRING POWER. DE-ENGERIZES ON A PULSE.

(Y)

NOT USED

LATCH RELAY

LOCKS ON SHORT TO GROUND

(X)

1

BREAKS VOLTAGE TO ABOVE RELAY WHEN ENERGIZED

RELEASE RELAY

2

POSITIVE VOLTAGE PULSE

192

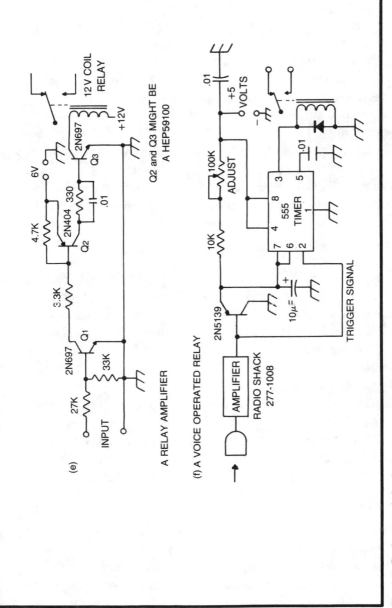

(e)

INPUT

27K

2N697 Q1

33K

3.3K

4.7K

6V

2N404 Q2

330

.01

2N697 Q3

12V COIL RELAY

+12V

Q2 and Q3 MIGHT BE A HEP59100

A RELAY AMPLIFIER

(f) A VOICE OPERATED RELAY

AMPLIFIER

RADIO SHACK 277-1008

2N5139

10μ=

10K

100K

ADJUST

.01

+5 VOLTS

4 8

7 3

6 555 TIMER

2 1

5

.01

TRIGGER SIGNAL

Fig. 6-5. Basic control circuits.

193

solid-state and integrated circuits. They love to piddle with these. So for them we will include some idea of how it might be used. We also recommend for them a book from our publisher entitled *Build Your Own Working Robot* by Heiserman, TAB book 841, which explored the solid-state concept in some detail and in a practical manner.

Let us fashion a relay tree as shown in Fig. 6-6. Notice that with a bank of 15 relays (or 4 relays with equivalent armatures and contacts) you can have 16 outputs of a common voltage, or make a common ground connection to the 16 output connections, one at a time. The routing is all accomplished by eight possible commands such as used by a computer. For example the command 10000000, where the digit 1 means a switch closure of S7 and none to the other seven switches on the diagram would mean the voltage from battery 1 would appear at output terminal 9. Trace this through for an idea of how all routing takes place. Closing S7 will latch relay A and no other. One more example. If we cause a momentary closure of switches S5 and S7, latching up A and the two parallel relays at C, we get an output at 13. To wipe out the latching we must momentarily energize switches S6 and S8.

Now this makes an interesting situation. If we were to connect all release switches together so that energizing one would energize all of them, we would have a single input which we could call clear. It would de-energize everything. We could then look at inputs to the four latching switches and relay them to a four-digit code such as 1000 to get an output at 9, and 1100 to get an output at 13, and so on. If you felt that the 16 outputs were not sufficient you could add another column of relays (16 in number). The number of output combinations you can have will be 2^5, or 32 outputs. You can, however, get an output at only one terminal at a time. In between the various operations performed by the robot having this brain, you clear the system by the closure of the common line to all release relays.

Figure 6-7 shows two circuits falling into the solid-state category, one using a multiple input to get an output, and another having an equivalent of the electrical latching system we just described. These circuits can be used but be aware they cannot handle the higher currents of the mechanical relays unless special transistors are used.

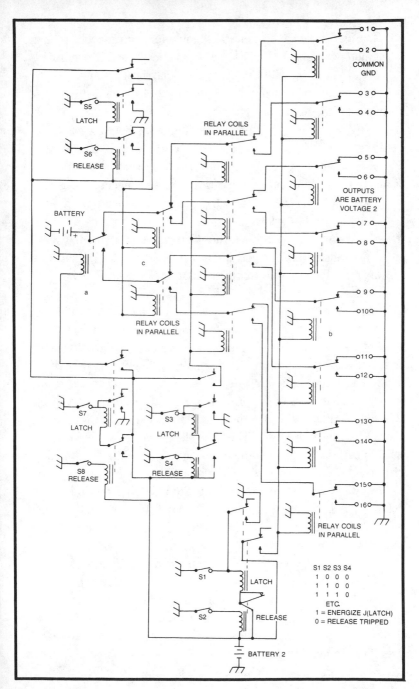

Fig. 6-6. A relay tree.

195

In Fig. 6-8 we have a general scheme for applying voltage to a drive motor in a robot. It uses the tree concept, a voice activated relay, and a light operated relay with a delay built into it.

We will assume that we have connected the drive motor to terminal 9 of the relay tree. Thus when relay A is latched, power will be applied to the drive motor and make the robot move forward. For clarity we have omitted all relays in between A and B, as shown in Fig. 6-8. Notice that the drive motor is energized through a set of contacts on a double-pole double-throw relay connected to give a reverse polarity to the motor when this relay is energized. This allows the robot to backup.

A signal comes in through the microphone, perhaps the word go. This causes the stepping relay activated by the microphone amplifier to step one movement to close the circuit to A. Relay A then latches up, and the robot will move forward.

Assume that the robot approaches a wall. Since it has a light and a light sensor on it, it picks up a strongly reflected signal which, through its relay amplifier, causes the delay relay to close. This, in turn, activates the motor reversing relay, making the robot back up and at the same time activates the steering section to make the wheels turn left from their normal neutral position. The delay holds this condition until the robot has backed away from the wall and is turned to a new direction. Then the delay releases, the steering goes to neutral due to a self-neutralizing feature and limit switching circuits, the drive motor reversing relay drops out, and the robot again moves forward. Now you want to stop the machine, so you say "stop." The microphone picks up this sound. It activates the stepping relay once again, this time activating the release relay of A and everything stops until you give the command "Go" again.

We have just used one of the commands possible to give the tree and, of course, another circuit for the light protection operation. We could have activated as many as 15 other commands with this simple brain system but only one at a time. You can easily trace things out to see why a trouble exists, and it gives you a way to go for your first rather rough robot. You can find out about the self-neutralizing and limit switches on the steering servo in our book *Model Radio Control*, TAB book 74.

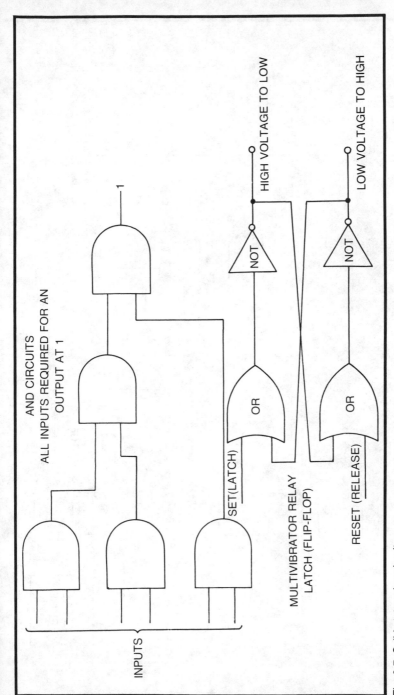

AND CIRCUITS
ALL INPUTS REQUIRED FOR AN
OUTPUT AT 1

INPUTS

SET(LATCH)

MULTIVIBRATOR RELAY
LATCH (FLIP-FLOP)

RESET (RELEASE)

OR

OR

NOT

NOT

HIGH VOLTAGE TO LOW

LOW VOLTAGE TO HIGH

Fig. 6-7. Solid-state relay circuits.

197

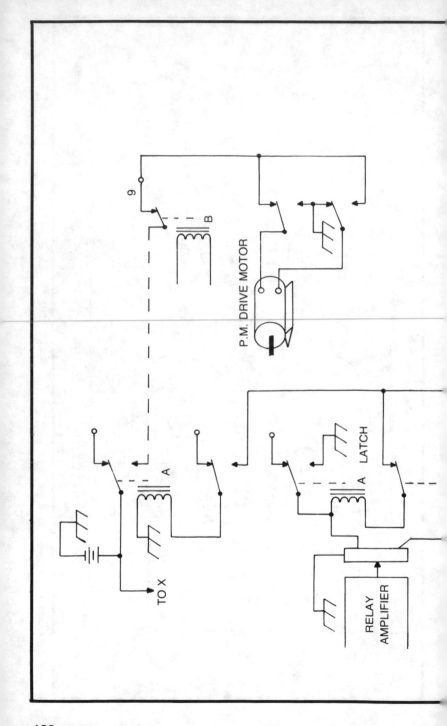

RELAY AMPLIFIER

P.M. DRIVE MOTOR

TO X

LATCH

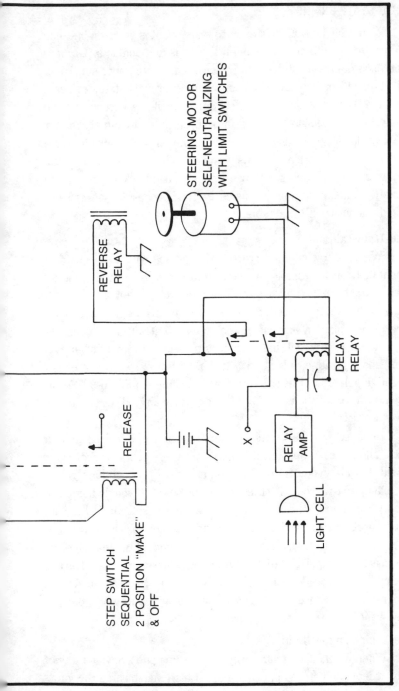

Fig. 6-8. Relay tree external circuit example.

199

SOLID-STATE BRAINS

With relays it is always easy to trace what is going on because you can follow the power through the coils and contacts. Granted that this is not the most desirable brain because we want a small one. If it is made from integrated circuits and other solid-state devices it will have the small size and excellent capability we require. And using the computer-like operation of integrated circuits allows our robot to do things which could not be done with the more massive and heavy relay brain. Let us take a good look at some basic computer circuitry and see how it might be used. We can also learn something about computer basics in this process if we are not already familiar with the subject.

The Truth Table

We need to examine a truth table and see what it tells us. You might be familiar with this but, if not, learning this method of routing signals might be helpful to you. The general symbology of a truth table is that a letter without a bar over it means a signal or connection. If we place a bar called (not) over that letter it means no connection or no signal. In some circuitry a letter without a bar over it means a positive pulse and a letter with a bar over it means a negative pulse. Look at the situation diagramed in Fig. 6-9. We can see how a truth table is set up for two relays, an AND circuit and an inverter circuit.

Because you can write connections in this special algebraic form, and since it is possible to manipulate this algebra in accord with its rules, you can determine the least number of connections (relays for example) to accomplish the circuit operation. If you do not follow the mathematical approach you can still have a satisfactory circuit, but it may be a little more expensive and complex than it really needs to be. There is literature available on the subject of Boolean algebra, which is the manipulation of this symbology. We won't expand on this further here. You do not need to understand Boolean algebra to understand the concepts which we will explain in this work, but you may want to learn it for your own use.

The Minicomputer Brain

When we think of a minicomputer we imagine a device that not only accepts and stores information but also manipulates the infor-

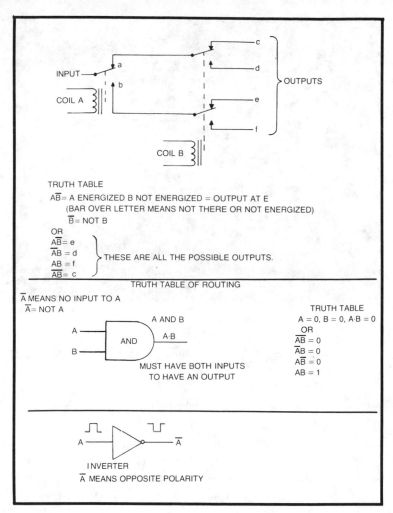

Fig. 6-9.Truth tables.

mation under control of instructions. It then presents the results on some output display or recording device. We also know that some computers are able to operate other devices in their outputs such as typewriters, control mechanisms for heating and cooling, and motors. These are called peripherals.

Because we can attach peripherals, especially motors, to a computer we now see in this unit the possibility of having the ultimate robot or android brain. We might even expect it to become so efficient that it can learn by doing. Then it can adjust its responses

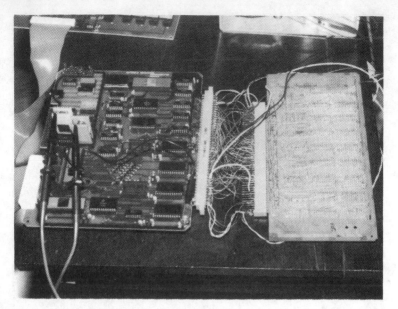

Fig. 6-10. A minicomputer exposed.

according to the latest information received from sensors. It might become almost "human!"

Well, just how far it can go to become self-sufficient isn't really known now. But we do know it takes a lot of circuits to do some things that, for example, a relay with multiple contacts can do quite simply. One type of minicomputer is shown in Fig. 6-10.

Yes, there are many thousands of circuits there but due to the small size of integrated circuitry, thousands of circuits can be placed in the volume of a single relay, so the actual number of circuits involved doesn't really bother us very much. What does become important is the speed of operation and the total capability, reliability, and freedom from maintenance. The solid-state devices perhaps excell here as there are no moving parts to worry about, nor sparks from relay contact closures or openings. And while a relay may operate in the thousandths of a second, the solid-state devices can operate in the millionths, even billionths, of a second. Some difference in speed isn't there? Heat is reduced and power supply requirements are smaller for solid-state devices too. Note the power supply for this minicomputer in Fig. 6-11.

Fig. 6-11. A minicomputer power supply.

Now we need to think about how we get information into such a computer brain. With this one a keyboard is used to program the unit. This is shown in Fig. 6-12 along with its interface cable. On this program keyboard are a series of integrated circuits which can generate test programs. And there is a series of numbers which light up to show the output from the computer when this is called for. This readout also can be used to check information going into the computer to be sure you have pushed the proper button as you do the programming operation.

This is also a kind of testing keyboard because you can designate certain memory locations and send them pulses to see if they actually accept and store these at the designated location. You find this out by giving the instruction to read out the information from that location. You push certain keys which give the computer this information, and then the lighted numbers tell you what was there.

To place that information into the computer at a certain location in the memory we had to have an address code. This opens the lines (gates) to that particular location where a flip-flop is set. Some computers use other methods of storing this information, such as a magnetic field. Later, when we call for this information from that address (circuit), the output lines will be opened by another instruction code so that only that particular memory circuit, or flip-flop, will present its voltage or absence of voltage to the common bus at that instant. This signal then is used to cause particular lights to light up showing what was held in that memory.

There are, as you realize, two states for a flip-flop bistable multivibrator circuit. If we have a voltage output this corresponds to a 1 in this binary code. A low voltage or zero output corresponds to a zero in the code. The combinations of ones and zeros can be large. A single voltage or zero is called a bit; a group of eight bits is usually called a byte, but any number of bits can be so called. It takes several bytes to make up a computer word. A byte might be written 11100000.

Because these circuits operate so fast, in the millionths of a second, it is impossible to check out the operation with an ordinary voltmeter. An oscilloscope—and a good one—with excellent response and wide bandwidth must be used. It is important that the scope leads be such that little loss is experienced from using them.

Fig. 6-12. A program keyboard.

So we see that if you plan to use a solid-state computer as your robot brain you will have a rather complex unit which must be programmed and tested with good test equipment. You will use the test equipment to follow through in its operation. The test program will be used to check the information you have placed into the robot brain. You will have to see if the robot responds to the programmed commands as it should.

You might use a keyboard to insert information into the robot or some other method. The ultimate is voice commands, of course, and just how that might be done is something to really think about. Well, let us move forward and now look at a simplified block diagram of a computer and see what might be contained in it from this viewpoint.

GENERAL COMPUTER OPERATION

There is so much literature available on computers and programming that you can advance your knowledge to almost any level you choose by studying it. But here we must take a moment to look at a block diagram of one and see how it works in order that we may proceed on a common ground in the pages that follow. If you are a computer hobbyist and a real expert in this field, we hope that your imagination will be stimulated. Perhaps you already know just how your hobby computer might be used to control a robot, but bear with us for the following examination.

The Computer Block Diagram

A computer must consist of several particular and distinct blocks as shown in Fig. 6-13.

The blocks are:

- A clock which furnishes pulses or pulse trains (bytes) at specific intervals of time.
- A data input device which may be a typewriter, a card reader, a keyboard, a microphone, or whatever. This produces bytes of information. You want the computer to use the information and from it determine an output.
- There must be a series of "gates" which are opened and closed by specific pulses from the clock bus. A bus is a common line to many circuits.

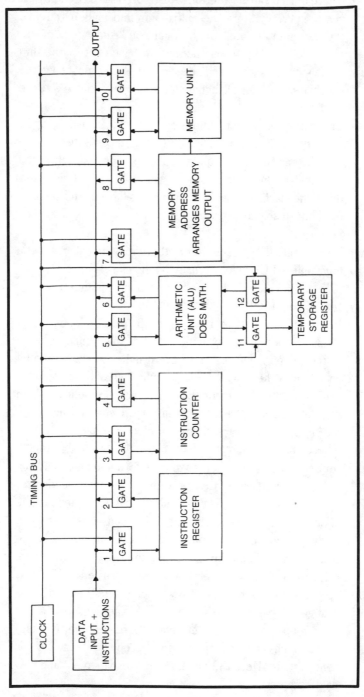

Fig. 6-13. Block diagram of a simple computer.

207

- There must be an instruction register. This is a circuit which can temporarily store the instruction information you give the computer. An instruction might be "Add two numbers given" followed by an instruction which says "store the sum in the storage register." The storage register must have the capability of holding all instructional information you give the computer.

- There must be an instruction counter, which will keep track of the number of instructions you give the computer and see that they are done in sequence, that is, one instruction at a time and in the proper order. This unit should also shut down the machine when all instructions have been executed or if the machine is given more instructions than it can perform.

- For computing type computers there must be an arithmetic logic unit (ALU) which actually opens and closes circuits to shift around electrical pulses to produce addition, subtraction, division and multiplication in the binary method. When it is necessary to temporarily store the result of one computation till another can be done, it sends this temporary sum to the temporary register where it is held a short time till called for by the ALU. When it is pulled out of the temporary register, the memory of that pulse arrangement is wiped out. All temporary register circuits are uncharged and ready for another temporary charge of electrical pulses.

- There must be a memory. This is the "biggie" in the computer because it is here that so much information is placed and may be kept permanently, or kept for the time needed for operations, or as long as you desire to have it remember the information. The memory may be thought of as a series of latching relays or flip-flops which can be set by the coded pulses coming in on the data input line from the bus. Some memory circuits have their information "burned in." That is, they are built so that once they receive data they cannot be changed, like turning a switch on or off and then sealing it so it cannot be moved once set. Then there are other memory types which can be changed when a proper code of pulses is sent to them. Here the old data may be wiped out and new data accepted.

Let us examine some types of memory units.

- The ROM (read only memory) has no means of obtaining new information. The data or instructions are placed into it at the factory. This is the sealed switch type of unit. To change the memory of this computer you must change the memory unit completely by wiring in a new one.
- The PROM (programmable read only memory). This can be programmed as you desire, but in the machine it can only be "read from" and not reprogrammed. Once programmed it keeps the program, but you put the program into it, not the factory. It's like a tape in operation.
- The EPROM (erasable programmed read-only memory). This can be programmed by you, and its memory can be erased and rewritten whenever you desire. It is like a tape which can be erased or wiped clean. But when it is in the computer it can read from it only and the machine cannot write into it.
- The RAM (random access memory). You can have the machine get into it for any information it has anywhere, at any time.
- The write-in/read-out memory. You can write a program into it or send data into it when in the machine. The machine can also write or read information into or out of it. This is like the EPROM but the computer has the capability of writing data into it when so instructed.

There may be other types of memories, but these will give you the basic idea. In the first category (ROM) it is interesting to note that toys can have such a memory in the form of specially shaped plastic cams. As springs cause these cams to rotate, the toy will perform certain functions and then do this over and over as long as the program is read from the cam.

In the form of the PROM is the light-switch timer, which has a means of setting lights to come on and off at any time or times during a 24 hour period. You can change the times, but once set, the machine will just execute your commands.

The EPROM might be like the magnetic tape which you can erase sections of and put the new data or information or instructions in the cleaned section. You might erase all information on the tape and start over. But you might have to take the tape out of the playback machine and put it on an erase and recording machine to put new data into it.

The write-in/read-out memory is one which can look at signals inside the computer and if it needs to change the arrangement of pulses in its memory because of what the computer is doing, or because of instructions you have given it, it will do so when a coded series of pulses come to it telling it to make that change. Sometimes this is called an adaptive system because it adapts to the current situation.

Realize that we have given only some examples of memory units, which are integrated circuits. There are many thousands of circuits in a single chip and many chips can be used, which then form the basis for the extremely large memories which we know even the smallest computers now have. Look again at Fig. 6-10, on the left, and see the circuit board with its neat array of integrated circuits. The probes here were attached for testing what was in some of the memories.

Back at our basic block diagram, we see that next to the memory is an address unit. The address unit designates which circuits in the memory will place their pulses on the bus when the memory out-gate opens.

The address unit is most important. It must be able to tap into every circuit in the memory and release its signal to the line *or* to open up that memory line for the receipt of signals if the unit is a write-in memory. The address unit will operate in accord with a specific code which you write on the input data card or tape, or perhaps type-in on a typewriter.

The Clock

Every computer must have a cycle of operation; that is, it must be able to connect whatever elements (blocks) are necessary for whatever purpose to the main bus line at definite and specified time intervals. For example, if we think of starting our block computer of Fig. 6-13, we want, during the first interval of time, the data input to be connected to the bus along with the memory unit so that data can be stored. It is also necessary for the address unit to be connected at this time so that the data can be stored in the proper place (specified by you in code) in the memory.

These three elements will be operated as follows: First the machine is turned on and the timing cycle started. During the first, say, two seconds (much shorter than that actually but we will use long time intervals in our explanatory example here), the input block, address block, and memory block are connected to the bus. Now by means of a code of pulses you cause the address unit to select a certain row in the memory which is to get the first information. Next, the clock opens gate 9 and your data flows into the memory to the row you have selected. The gate then closes and the memory unit is disconnected momentarily from the bus. The clock keeps ticking and now causes the address unit to become disconnected and causes the instruction register and program counter to be connected to the bus. When the clock opens gate 1, information and instructions as to what to do with the information in the memory are sent into this block to be held in temporary storage. Then the program counter is given information as to what to do to accomplish these kinds of instructions as steps. For example, Step 1 would be "Fetch the data from the memory and put it into the arithmetic logic unit." Step 2 would be, "Fetch another number from the memory and send it to the ALU". Step 3 would be "Add the two numbers and temporarily store the result, through gate 11." Step 4 would be, "add the result to the next data from the memory at address x", and so forth.

The clock opens and closes the various gates for a specified time interval until all necessary operations can be performed and the final operation of placing the result on the bus, to be picked off at the output, is accomplished. Then the whole timing cycle is repeated. It is repeated over and over until all input data and all instructions are used up—that is, all the instructions and calculations have been performed and all the results of the operations are present on some display in the output. Then the machine shuts off. The clock times everything so that everything happens in a logical order and nothing gets confused or mixed up.

Instruction Codes

The instruction code that you give the machine actually may be given right along with the data and it tells the machine what to do with data pulses. It tells it which ones to store in latch-up circuits for permanent retention, which ones to store temporarily, and which ones to use immediately. This instruction code is called the language of the computer. It has such names as COBOL, BASIC, and machine language. You have to use the code which *your* machine understands in order to make it do what you want it to do.

THE ROBOT BRAIN CONCEPT

So, we might ask ourselves, what does this all mean to a robot master? It means that in designing a machine he has a choice of using this system, or one like it, to make his robot perform tasks and respond to various signals and stimuli. Instead of a simple display output which might represent some calculations, the robot master will usually expect some physical action, or some response other than just a display. The physical actions may be the results of some calculations and comparisons (of input and output) within the brain section itself.

Now let us examine Fig. 6-14 wherein we have indicated one basic method of how a robot's brain might be arranged using computer techniques. You computer hobbyists will have a ball making such a diagram as this into a real working model computer.

We have on the left a series of sensors which are scanned very fast by a sampling action so that each sensor's data can be presented to bus 1 immediately to the right. The scanner also scans the outputs at the same time and in synchronization so that if some input data needs to go to a particular output device, that output device will be so connected to receive it and won't get data which should go somewhere else.

Bus 1 has several output connections, one to a memory which has stored in it various responses (actions) should these be initiated by some input. An address unit sees to it that only the proper memorized response can go to a proper output device so you won't have a talking response trying to get through to a steering motor!

The memory unit also feeds a comparer unit which can, through the output bus 2 and related gates control each of the many segments of the robot, such as the head, voice, eyes, various arm movements, and general movements. Thus this unit not only compares but directs actions as well. Notice that we have indicated, but not directly, that each output device must have the proper amplifiers, relays, and other items to make the specified output do what it is commanded to do.

There is a block called clock, scanner, and timer. We have indicated what these do in controlling the timing and scanning actions of the inputs, outputs, memory, and other units. The clock would be started when you command the robot to "Wake up." It would keep timing the operations until you voice another command such as "Go back to sleep."

In the diagram is a block called the direct action initiation. This has a direct connection to the input and output bus. This unit is directly responsive to certain sensors. If it gets a signal from them, and is made responsive by a signal from the address unit, then it has the circuitry to override anything that the robot may be doing at that instant. A case in point might be if the robot were moving and suddenly its light sensor got a strong reflection from a wall. Recall that we discussed this earlier? Then this direct-action unit would have to stop the robot immediately and make it move backward and change direction. This circuit would have to disconnect the memory block then insert its own particular program in its place to avert damage or danger to the machine. Much thinking can be put into just what and how this direct-action block should do and operate.

The memory or program section of the brain might be a floppy disc which with instructions written on it or a tape or some equivalent. The memory should be a RAM type wherein the machine can get to any program quickly and accurately within split seconds. Tapes might be used for voice, but whirling a tape back and forth inside a moving robot, hunting a particular section or segment, might take too long for the responses we want. It might require too much machinery for the purpose too. A small floppy disc memory might be just the thing. With them it is easy to change the program content (with a new disc). And they can quickly position the readout head to any track on the disc to get any particular program or response. They are also relatively small. Of course you might use other type memories if you do not have too many permanent programs such as shaped cams and timer switches. But you will need a means to insert data into them (or replace them) and wipe them clean of old data or programs when necessary and do this easily and

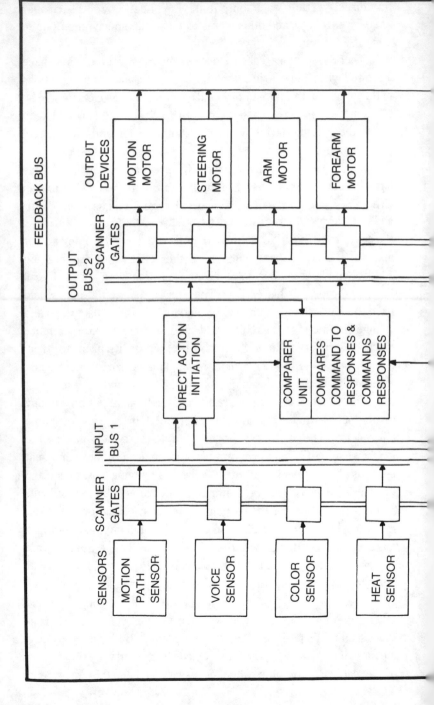

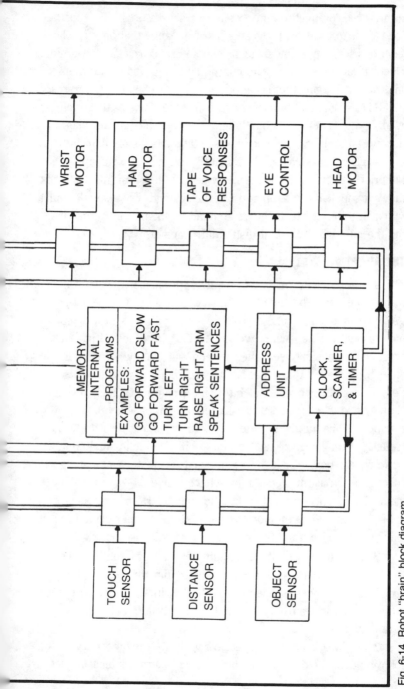

Fig. 6-14. Robot "brain" block diagram.

215

accurately. You might want to adjust just one program. You then don't want to erase all programs just to change that one. Replacement of discs, tapes, and cams are some ways to change programs. Electronically, dissipating the charge on integrated circuit memory units and recharging them is another means of changing programs.

So this block diagram is to give you some concepts that might be used. If you want to start blocking out your own computer diagram, start with a big block showing inputs and then begin breaking it down into smaller and smaller blocks until you finally wind up with a blueprint of circuits and/or devices which do all that the master block shows it should do. It's fun and a real challenge. For example, start with a large block which shows an input of "Signals from the sink," and the output is "Washed dishes into cabinet."

THE ADAPTIVE SYSTEM

This robot system most closely approaches what we, as humans, consider to be the ideal. In fact, this is the way we operate. We have some learning and experience which is stored in our memory cells. Then when we are faced with a new situation our brain scanner quickly searches throughout our memory banks for anything and everything that has any relationship to this new situation. Then piecing together all bits of information that might be useful allows us to take a calculated-risk movement or action in that new situation. Once we have done this, our brain analyzes the results (actually it may analyze as we start doing the action and continue while we are doing it). If the results are satisfactory—meaning the absence of an alarm signal somewhere inside us—it causes us to memorize that action for that situation and a new program is stored in our memory. The memory will also store the information that it didn't work, so we won't try that solution for that problem again.

So the ultimate robot brain might be an adaptive one. This brain would have a trial and error section to function when all other programs or responses failed to satisfy the situation confronting the robot. This can also be called a learning robot. It learns by experimentation.

There was an adaptive system built by a large university. In it a robot mouse was to go through a maze of passages built in such a

216

manner that the mouse could not get through them directly by chance or accident but might wander back and forth, round and round, and never get through if there wasn't some guiding intelligence and memory to help direct it.

The mouse robot's actions were kept track of in a computer memory as to direction and length of movement and success or failure of advancement. Then the experiment was started. At first the mouse did wander, trying all avenues of passage. The computer kept track of where the mouse was all the time and what it had done to get there (which direction it had gone, which passages it went through). In time the mechanical mouse had gone through every possible passageway in the maze and finally had come to the exit.

Meantime the computer had been "learning" by memorizing those passageways which gave movement through the maze and those which did not result in forward progress. In a quick calculation (for the computer), the brain determined which routes the mouse should take to get through the maze quickly and directly. The wizards then placed the mouse at the entrance of the maze immediately without any problems or delays of any kind. The route through the maze had been "learned" by the computer brain.

So we envision that our robot might learn to navigate our homes and yards and our walks and streets if it has a computer brain to analyze where it had been, how far it had come, and what obstacles it encountered in its path. It must be taken through the course the first time. Then it calculates the most direct path to get by any obstacles. The tiny size of our currently great capability computing equipment now makes this concept possible and not just a dream as it was a few years back. An adaptive robot closely approaching the android is a possible reality for those of us who can fabricate this brain for it.

Beyond the mere movement learning process, the robot might learn other tasks. That is, if it can't do it one way it will learn how to do it another way. For example in speaking if it doesn't get the proper response to its statements or questions it may learn to rephrase the words to get a better response. It might also learn to threaten you if you didn't come up with sensible and good responses—or could it?

Modern industrial robots are guided through tasks and processes the first time around by a human controller. The computer memory learns the magnitude of control signals which are thus generated, their size, polarities, and duration. It also learns the sizes, polarities, and durations, of feedback signals needed when the robot is doing exactly what it is supposed to do. Then the computer can generate those exact control signals and look for the exact feedback signal which makes the robot precisely duplicate the actions which were human guided the first time. Thus a robot "learns."

Well, so much for this subject at this moment. We need to look at sensors again. As you can see we need to have many varieties. These are really what governs just how much our robot can do. If it can't sense it, it can't do it. Actually the sensors may be the keys to our whole autonomous robot operation.

CHAPTER 7
AN ADVANCED LOOK AT SENSORS

We considered sensors to some extent in an earlier chapter, but since these are the devices which will enable our robot to learn of the world around it and how to react in that world, we need to exmamine them further at this time. We will be examining more advanced sensors and their use in complex systems. We will also be examining what sensors will and won't do and what they should and should not do in operation. So if you border on being a wizard you might find this interesting, and if not, you can still gain much information from a study of this chapter.

A FORCE FEEDBACK SENSOR SYSTEM

In a study performed by MIT for the Office of Naval Research there is a force feedback system described for a robot which might be called upon to replace humans in some tasks. This study, called "The Little Robot System," was done by D. Silver. We have taken some liberty in expanding some of the system explanations in an attempt to clarify the ideas for a more general audience, but the basic concepts, symbology, and mathematics are as stated in this report and we have made some extracts from it.

The purpose of the study was to investigate the possibility that such a machine could, in fact, perform delicate and demanding tasks in assembly and construction, tasks which would ordinarily require a human who was well trained and skillful in the job.

Here we are particularly interested in how the force feedback sensor system was conceived, and according to its equations, how it would give information for control of a hand and wrist, perhaps an arm also. We do realize that the concept of this robot was of a machine fixed in a particular location, so it could have a particular set of coordinates related to its body position. All other forces, movements, and angles could then be referred to that set of coordinates by its computer brain. Here is an extract from the study, modified to some extent in places in an attempt to simplify the explanations.

THE LITTLE ROBOT SYSTEM

As mentioned, the Little Robot System was developed by D. Silver. It is a medium size, 5 degree of freedom, seven-axis robot which is controlled through a computer and a programming language called LISP. The computer is a PDP-6 type.

At Fig. 7-1(a) we see a sketch of a robot and its arm, included to pictorially show the two basic coordinate systems involved. The robot has its own set of axes. You see the x axis running through the shoulder, center of gravity, of the robot and the x" axis, an axis of the force sensor complex, located at the wrist.

At (b) you see a diagram showing six linear variable differential transformers located on the sensor axes and at angles to these axes. Note angle (a), which is an angle between the x axis and the x" axis. The robot coordinate system is related to the force sensor complex coordinate system through this angle as subsequent equations will show. Since we are interested in torques, which are forces times distances, we need the force value sensed and the distance from the wrist coordinate axis, which is the value d.

It is interesting to us that there are six force sensors located in this wrist. We think of the complex arrangement of bones, muscles, and tendons in our own wrists and know that this part of our body is one of the most amazing we have as far as motion capability is concerned. It turns and twists and bends in all directions. So to duplicate its motion we must have a machine able to make a hand do complex manipulations. If we are to replace a human in a task, as Silver did, the exploration of very complex motions must be undertaken.

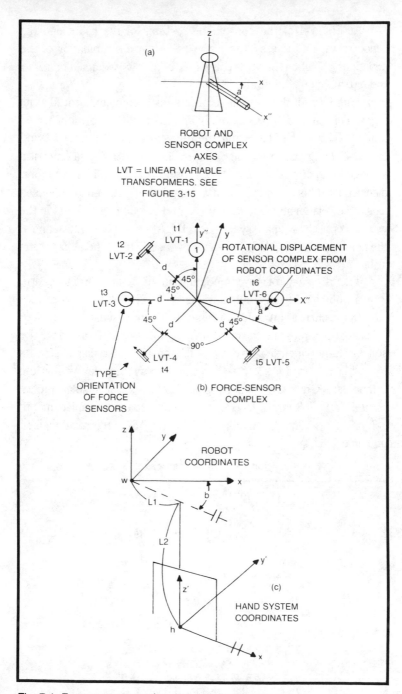

Fig. 7-1. Force-sensor complex and coordinate system.

At Fig. 7-1(c) the coordinate systems of the robot and the sensor-complex force-system appears. A close examination of the equations will show how they are related. Now, we again consider the extract from the study.

The heart of the robot is a force sensor complex located at its wrist. This complex consists of six linear variable differential transformers (L.V.D.T's) to permit the measurement of the forces and torques acting on the wrist. Figure 7-1 shows the geometrical arrangement of the six L.V.D.T's, where angle a is a rotational displacement between the robot axis and the force-sensor-complex axis. The relationship between force and torque components in the w-xyz (robot) coordinate system and the force components acting at the position of each L.V.D.T sensor is described by the equations in Table 7-1.

The force and torque components in the *hand* coordinates are, in turn, obtained from the equations of Table 7-2.

In the current Little Robot System, an approximation is used to calculate the forces and torques, since the stress-strain matrix of the sensor has not been analyzed in detail. Consider small displacements (d_x'', d_y'', d_z'') in the x'', y'', z'' coordinate system, as well as the rotations (d_a'', d_b'', d_c'') about the x'', y'', z'' axes. The displacements in the L.V.D.T.'s $(t_1, t_2, t_3, t_4, t_5$ and $t_6)$ are related to displacements d_x'', d_y'' and d_z'' and the angles d_a'', d_b'' and d_c'' as in the relationships of Table 7-3.

Table 7-1. Force Equations

$$F_x = -f_2 \cos(45 + a) - f_4 \cos(45 - a) + f_5 \cos(45 + a)$$

$$F_y = -f_2 \sin(45 + a) + f_4 \sin(45 - a) + f_5 \sin(45 + a)$$

$$F_z = f_1 + f_3 + f_6$$

$$T_x = f_1 \cos(a) - f_3 d \sin(a) + f_6 d \sin(a)$$

$$T_y = f_1 d \sin(a) + f_3 d \cos(a) - f_6 d \cos(a)$$

$$T_z = f_2 d - f_4 d + f_5 d$$

where F_x, F_y, F_z = forces in x, y, z axes
T_x, T_y, T_z = torques about x, y, z axes
$f_1, f_2...f_6$ = forces at each L.V.D.T.

Table 7-2. Hand Coodinates Force and Torque Equations

$$F_x' = F_x \cos (b) - F_y \sin (b)$$
$$F_y' = F_x \sin (b) + F_y \cos (b)$$
$$F_z' = F_z$$
$$T_x' = ((T_x - L_2) \, F_y) \cos (b) - ((T_y + L_2) \, F_x) \sin (b)$$
$$T_y' = ((T_x - L_2 \, F_y) \sin(b) + ((T_y + L_2) \, F_x) \; (\cos(b) + L_1) \, F_z$$
$$T_x' = T_z - L_1 \; (F_y \cos (b) + F_x \sin (b))$$

By the coordinate transformation from the sensor axes (x'', y'', z'') to the robot axes (x, y, z,) we can obtain the displacements (d_x, d_y, d_z) in the x, y, z, coordinate system, as well as rotations (d_a, d_b, d_c) about the x, y, and z axes. Next we must determine the overall stiffness (or servo gains) *in* x, y, and z axes and *about* the x, y, and z axes (meaning linear and angular gains). Then we can get force and torque components *approximately* as related in Table 7-4.

The servo control of the robot is done by the PDP-6 computer every 1/60 second. LISP in a PDP-10 govern the PDP-6 via five special LISP functions, as listed in Table 7-5.

The PDP-6 variables are shown in Table 7-6. The prefixes X, Y, Z, R, G, V, T mean X-axis, Y-axis, Z-axis, rotation about Z, grip,

Table 7-3. L.V.D.T. Displacements

$$d_x'' = \; - \; \frac{(t_2 + t_4)}{\sqrt{2}}$$

$$d_y'' = \; \frac{(t_4 + t_5)}{\sqrt{2}}$$

$$d_z'' = \; \frac{(t_3 + t_6)}{2}$$

$$d_a'' = \; \frac{(t_3 + t_6 - 2 \, t_1)}{2}.$$

$$d_b'' = \; \frac{(t_3 - t_6)}{2}$$

$$d_c'' = \; \frac{(t_2 + t_5)}{2}$$

Table 7-4. Force and Torque Components

$$F_x = k_1\, d_x$$
$$F_y = k_2\, d_y$$
$$F_z = k_3\, d_z$$
$$T_x = k_4\, d_a$$
$$T_y = k_5\, d_b$$
$$T_z = k_6\, d_c$$

vice, and Tilting-part-holder, respectively. Mode switches FRE select the control mode. For example, when FRE is 0, the X-axis acts as a position control servo, and when FRE is −1 it acts as a force control servo.

What follows is a brief introduction to the basic functions used in a study of the Little Robot System.

- *Position control (absolute)*: X=, Y=, Z=, R=, G=, V=, P= eg: (DEFUN X=(X) (SETM XDES X) (SETM XFRE 0))
- *Position control (relative)*: DX=, DY=, DZ=, DR=, DG=, DV=, *DP* = eg: (DEFUN DX= (DX) (X=(PLUS (GETMF XPOS) DX)))
- *Force Control*: FX=, FY=, FZ=, FR=, FG=, FV= eg: (DEFUN FX= (FX) (SETM XFDS FX) (SETM XFRE −1))
- *Check of Position Control*: ?X, ?Y, ?Z, ?R, ?G, ?V, ?P eg: (DEFUN ?X — (SEO (GETM XDLT) 0 threshold-P))
- *Check of Force Control*: ?FX, ?FY, ?FZ, ?FR, :FG, ?FV eg: (DEFUN ?FX — (SEO (GETM XFRS) (GETM XFDS) Threshold −F))

Table 7-5. LISP Functions

SETM	(SET mini) of two arguments, sets PDP-6 variables
GETMF	(GET mini floating point) of one argument, reads PDP-6 variables in floating point modes
GETM	(GET mini fixed point) reads PDP-6 variables in fixed point mode
WAIT	of one argument, evaluates its argument every 1/30 second and will "sleep" until evaluation returns T
SEO	(Sort of Equal) of three arguments, returns T when the first argument is equal to the second within the tolerance of the third, otherwise NIL.

- *Miscellaneous*: X0F, Y0F, Z0F, D0F and H0F gets X, Y, Z coordinate. Height and diameter of object is specified by the angles (argument), respectively.

Position control functions as well as force control functions deliver their arguments (angles) to the PDP-6 when they are evaluated. From that moment the servoing loop in the PDP-6 computer controls the robot to the specified destination, in the specified mode, until another function changes the control. The function WAIT *checks a control situation* and waits until its argument (angle) evaluates to T. The servoing is still in effect after WAIT is terminated.

An example:

(FZ = fz) (WAIT ' (?FZ)) (DR = 3.14) WAIT ' (?R))

This program moves the hand down until it lands on the table, i.e., a *force is felt in the Z direction*. When (WAIT ' (?FZ)) returns control, the internal program proceeds, and rotates the hand the specified 3.14 radians. During, and even after this rotation, the hand keeps the contact force fz against the table, because (FZ = fz) is still in effect. Then the robot would continue with the continuation of its program.

Table 7-6. Inputs and Outputs

INPUTS				
suffix	prefix	meaning	unit	type
POS	X, Y, Z, R, G, V, T	current position	inch/rad	floating
DLT	X, Y, Z, R, G, V, T	position error	pot. unit	fixed
FRS	X, Y, Z, G, V,	calibrated force	L.V.D.T. unit	fixed
TRC	X, Y, Z	calibrated torque	L.V.D.T. unit	fixed
OUTPUTS				
suffix	prefix	meaning	unit	type
DES	X, Y, Z, R, G, V, T	position destination	inch/rad	floating
FDS	X, Y, Z, R, G, V	force designation	L.V.D.T. unit	fixed
FRE	X, Y, Z, R, G, V,	mode switch		0 or -1
GAN	X, Y, Z, R, G, V, T	position servo gain		fixed
FGN	X, Y, Z, R, G, V	force servo gain		fixed

We explained some parts of this in more detail than the original author did in his study. We did this to make the explanation easier for understanding at the level of this book.

It is also interesting, we think, that when designing a working robot, one can mathematically specify equations to be solved by the machine as it moves its head, arms, wrists, hands, and legs. The equations are called algorithms. If you can work out all the equations of motion, you then know what variables will be important, such as power, linkages, friction, forces, angles, distances, speed of motion and so on. In the solutions, of the equations you find the proper quantities to use. The computer associated with the machine can then use proper sensor feedbacks to make the machine do exactly what you want it to do.

We included this excerpt so that you computer bugs can have a go at the mental exercise of analyzing the forces and quantities as specified by the equations of the Little Robot. For those of you who might want to avoid such a task, just remember that force measuring sensors are used. They are mounted at the wrist and measure forces on the wrist and hand from various directions. These forces convert to feedback signals which are sent to the computer directing the wrist and hand.

When the hand is in the proper place to grasp an object it can feel forces on its hand in various directions. So it grasps the object with a certain tightness and completeness of grip due to its accurate position on the object. The L.V.D.T.'s mentioned might be the type of sensors referred to in Fig. 3-15 if there is, for example, some spring arrangement on the movable part of the core "leaf" so that it takes a certain force, or torque, to cause it to increase or decrease a signal which it generates.

OPTICAL SCANNERS

An optical scanner is a sensor related to the human eye and can take many forms. If we use examples from radar engineering, we can devise sensors to give us information defining objects by their lightness or darkness against some background. Our robot might be said to see these if it has photoelectric eyes. The light and dark information might be used with a photo negative. By linearly scan-

226

ning the light streaming through the negative and comparing it to light reflections it sees when also linearly scanning the area ahead, the robot might determine a path to follow. Following this line of reasoning, then, we could insert a strip of film in the robot. This could automatically be advanced from one frame to another, each time making a comparison of the light and dark, and even gray, areas. Each time the robot would move in an attempt to match up each frame.

Photoelectric sensors are most important in hand-to-object situations. If a hand has a light so arranged that it beams into many light cells in the hand, then when this beam is interrupted by the hand moving over an object, for example, the hand might be caused to close on the object. Thus the light beam interruption tells the robot control brain when it is near or far or has inside it some object of concern. Notice that we are implying light scanning sensors on the outside of the hand as well as in the grasp area.

A Light Scanning Process

We include a typical scanning arrangement because it is a proven method and also might apply to infrared, visible light, radar, TV, or other methods you might use to detect objects near it.

If we use an off-center tube arrangement so the tube and some lenses restrict the field of vision of an electric eye, we can cause this tube to rotate in an off-axis manner so it will scan a cone-shaped area around and including its axis. It can look ahead of the robot to the side, or to wherever the robot needs to look by proper adjustment of its scanning mechanism. The electric eye can produce a voltage through proper circuitry which is exactly proportional to the light from each segment of the scanned area. Figure 7-2 shows the concept in rough form.

The output voltage of graph (a) would have a form similar to a positive going DC wave with its maximum in the angular direction where the most light exists (90°) and the smallest voltage where the least light exists (270°), according to how we assumed the light situation here.

To determine the direction of maximum light, a reference voltage is required. Then the cell voltage can be compared to that of the reference voltage and the direction learned. Its use in the robot would be automatic as, for example, if the reference voltage leads the signals voltage, as indicated at (b). This might call for the robot to turn right. If the reference voltage lagged the cell voltage this might mean the robot should turn left. Thus left and right motion could be governed by this system.

THE SCANNING REFERENCE

As in a radar system the reference generator is on the shaft of the scanner itself. This produces an AC voltage which is phase related to the shaft orientation as in Figure 7-2(b). This voltage is always fixed in phase and magnitude relative to the shaft position. If the signal from the eye is compared against this signal, the resultant voltage tells the robot which direction to move to zero out the incoming voltage by having it directly out of phase with the reference voltage, or it could move to have it exactly in phase with the reference voltage.

There are many integrated circuit voltage comparison units which will give a positive DC output for an out-of-phase condition and a negative DC output for an in-phase condition of the two voltages fed into its input. These output voltages could be used to control a steering motor in the robot.

This type scanner can apply to many variations of sensors. There may have to be some modifications to the concept to adapt it to your own sensors. You might even consider a small radar sensor using this concept.

TACTILE SENSORS

No well-constructed robot would be complete without a sense of touch. This is related to a human's skin sensing when he has hold of objects or has his hands, arm, or body in contact with something. Touch is very important.

In a robot system the sense of touch can be accomplished by having tiny microswitches with "cat whisker" feelers extending a short distance from leaf-type operating levers of the switches. Or you might have a plate attached to a small movable lever which can close the switch with a very tiny movement. Some of these micro-switches require only a few thousandths of an inch activation of the lever to trip them, yet they give good positive action of large

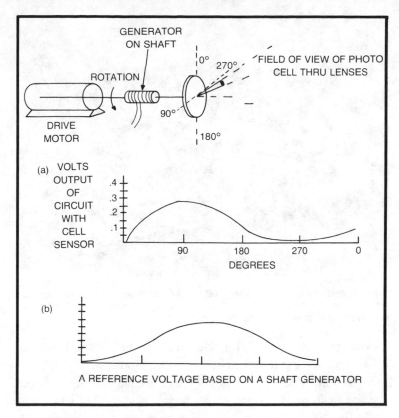

Fig. 7-2. Basics of an optical scanner.

contacts able to carry required signals to the brain of the robot or to other control circuit devices such as motors. If the signal goes to the brain a flip-flop circuit can be activated to cause motion of a hand, arm, or wrist to stop when it comes in contact with anything. This can give a sense of touch to a robot so that it can tell when it is at a wall or other obstruction. The signal might cause it to turn away. Let us examine an example.

First the robot moves toward the wall while holding its arm out slightly. It might be partially bent or extended fully. Touch sensor would be on its hands. When its touch sensor first contacts the wall, it immediately causes the motor drive circuits to open, and the robot stops moving. Then its other arm is caused (by internal programming) to move up and verify the size and shape of the obstruction. If for example, it is a wall—verified by the fact that throughout the

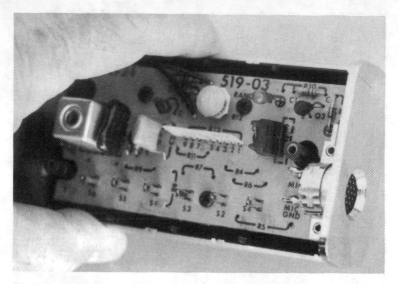

Fig. 7-3. A sound generator. The small output transducer to the right radiates sound above the hearing range of humans. Various tones are transmitted by pushing control buttons for desired operations.

movement of the second arm sideways, up, and down, its sensor tells it something is there—then this second touch signal combined with the first (which is still there because of contact with the wall of the first hand) could cause another internal program to be initiated to cause the robot to back up, say a foot or so, and turn through a 90 degree turn and try moving forward again. Touch sensors like these have many uses and should not be ignored on your robot.

PROXIMITY SENSORS

There are many varieties of proximity sensors to incorporate into our robot. One type of proximity sensor is the sound-echo system (Fig. 7-3). Similar to it is radar, light reflection, and heat sensors (infrared). There are many such devices on the market shelf which use these methods in home security systems. These might be adapted as proximity sensors in our robot. It is said that a robot having proximity sensors like these can move at a relatively high speed without danger of a collision. It might also search for objects till it finds them—by a signal from the proximity sensor.

Here is one application where a low-powered laser could beam a pinpoint of light for triangulation to determine not only that an object

230

is there but also the distance to it as compared to the distance to something else, like a wall. It thus determines that there *is* an object in the room because of the range difference to it and its surroundings. Circuits to measure the time that it takes to send and get back the light echo (reflection) are like those used in radar systems. But remember to be careful of your eyes with lasers!

COLOR SENSORS

Because you can use color filters with light-activated cells you can give your robot the ability to identify colors and thus objects which might have these colors. By comparison of the outputs of various cells with color filters, the robot can see which reflection most nearly matches, for example, a prerecorded memory voltage of that color as sensed by his sensors. This way he knows what colors are around him. Thus it seems possible to equip our robot with a means of finding objects by their color as well as their shape, size, and proximity and distance.

TELEVISION SENSORS

The use of a regular TV camera might be in order if you want to control your robot by radio or cable commands from some distance away. Figure 7-4 shows how TV cameras with their illuminating lights are used in the Hughes Aircraft Mobot Mark II system. Notice also that one "hand" of this remarkable machine is holding a geiger counter to test for radiation. With the arms, wrists, and hands used the duplication of human arm, wrist, and hand actions is quite good. One type control console for such a machine is shown in Fig. 7-5.

USE OF FIBER OPTICS IN SENSING

The use of light emitting diodes (LEDs) makes fiber-optic sensing possible. They may be located back in the arm or hand at some convenient location and the light emitted by them carried to the hand through bundles of fiber-optic material. You've seen these bundles on displays in light stores where the tips glow from light carried from within a central container.

In proximity systems the light is projected downward in two beams. The radiation reflected from the object will come back through a parallel series of fiber-optic conductors and go to light

Fig. 7-4. The Mobot Mark II. Courtesy Hughes Aircraft.

activated cells. The signals controlled by these light activated cells will then be sent as feedback information to the comparison amplifier producing the hand closing signal. This proximity system is being investigated by the National Bureau of Standards. They would be a source for further information if you should be interested in obtaining it.

The manner in which this system is used is to compare the voltage output due to reflected light from each of the two pairs of return light channels, one on each side of the hand grip like on the middle finger and thumb of a human hand. If the signals are identical then the object from which the light is reflected is exactly in the center of the grip; if the signals are not the same, then by knowing the larger voltage a part of the robot's brain, a comparison amplifier,

Fig. 7-5. A control panel for the Mobot Mark II. Courtesy Hughes Aircraft.

can determine which way to move the hand to center the object in it. Note that this is a short range position determining system, being used only for ranges in inches.

With the use of many fiber-optic channels it is possible to provide a hand with a position sensor for many directions. Even the back of the hand might be so instrumented. Then when the robot starts looking for an object it is to grasp, it can determine whether the object is *in* the grasping fingers, *outside* of them, or on one side or another, or ahead or behind the hand.

This gives a kind of "sight" to the robot right at its hand position so, along with touch sensors, the hand almost becomes a robot in itself. The idea of using LEDs, fiber optics, and light sensitive diodes (resistor-types) allows one to visualize many advantages for this system as well as to stir the imagination to other possible uses of this instrumentation.

HEAT SENSORS

It is possible, in the hand of a robot, to use infrared radiation to determine the proximity of hot objects and thus be able to move the hand to a predetermined position with respect to the object so that it may be grasped firmly and precisely.

Some kind of fast acting thermometer, spring, or perhaps a cell which responds to heat radiations in the infrared region might be used for this element. It could provide a robot with a fire warning or alarm capability, which you might admit will be a useful thing to have in the home. If our robot is to be a security guard for the house and patrol certain areas at night on a prescheduled basis then it would be important to have the robot sensitive to heat. Then it can alert us if there was a fire, or any undue heat situation in this area. Note that this is not the same as a smoke sensor used with most current fire warning systems, although that could be used.

We have indicated that the temperature sensor should be fast acting, which means within a reasonably short time it can give a contact or voltage output. Some thermometers are very slow and would not give sufficient output before a fire could consume them! Some expanding spring type temperature gauges are also slow to react. But it is possible one might use the conventional principles of temperature measurement and refine the instrument so that it works faster, either by using special metals or shapes in coil springs or a special channel in the mercury type thermometer temperature instrument. There may be other ways to detect temperature variations on a large scale also.

If we consider the ultimate, we could perhaps visualize our robot having a temperature sensitivity such that it could determine whether an object near it was a human or similar life form—such as some other mammal. These all radiate heat, and at close ranges might be found to produce sufficient energy output to make them detectable. Some insects use feelers for this purpose, and it is said that a snake can detect fauna life forms through its sensitive tongue.

SENSORS SIMULATING THE HUMAN HAND GRIP

There has been much work done at various technical and scientific institutions on the construction of hands and arms which can be attached to the human body and activated by electric currents produced by the body. The "Six Million Dollar Man" TV series was one example of a flight of fancy using these kinds of devices.

But they are real and science is advancing concerning their use. We show in Fig. 7-6 a concept of such a hand and its controlling mechanism as experimented with in Russia. This device is important to us as we consider the development of our robot into an android. (I do hate to use the word machine concerning these things. They are getting more human all the time. And with the capabilities associated with them you may get to love them after all. By the way, they are much more affectionate than the Pet Rocks were!) The device presented in Fig. 7-6 reacts to a load and is a closed electromechanical system with feedback exactly reproducing certain characteristic functions of the human hand.

In contrast to ordinary servo mechanisms this system has these characteristic features:

- Its feedback operates with respect to maximum error.
- The system is of an indeterminate nature, and the same terminal state can be achieved by different paths.
- The error signal is caused by deviation of the hand from the initial normal state.
- The system performs the two most characteristic reflex actions of the human hand: It holds an object, and it adjusts the pressure to the weight of the object.

The structure of the investigated device is basically identical to the anatomical structure of the human hand. The holding device adapts automatically to the shape of the object in it and determines whether grasping or pinching motion is required. The force of compression on the object is regulated automatically according to its weight. The inside surface of the holding device is sensitive to the pressure applied to it. (Signal increases as carbon granules are compressed.) For this purpose a number of holes filled with powdered carbon, which is usually used in microphones, is made into the sheet of rubber forming the surface of the clamp support. The openings on both ends are closed by a current conducting metal surface. The current created by the elastic pressure-sensitive surface is proportional to the load pressure and the area of its application. (Note that the conducting surfaces must also be flexible.) The surface of the clamp is the source of the yes-no signals. The signal fixes the time of contact of the object with the clamp and causes the compression reaction. In order to obtain sufficiently good sensitivity, rubber with low elasticity is used. In addition, the surface of the clamp (hand) generates signals proportional to the applied pressure. Observation of exact proportionality is not required here. The investigated source of control signals essentially increases the possibilities of the devices.

The servoamplifier has four inputs:

1. Potentiometer P1 is used to establish the initial conditions. It is placed on the forearm so that the operator can control the position of the clamp by pressing the potentiometer to the body.

2. At the time the clamp presses against any object the surface, which is sensitive to pressure, begins to send out control signals. If these signals exceed the threshold level, on the basis of the positive feedback effect, the clamp begins to press together. As a result of separation of the clamp (hand) surface into two zones, it becomes possible to determine which of the two motions—grasping or pinching—is required. The pressure ends of the feelers (fingers) causes a current in one direction (polarity) at the same time as the pressure on the surface of the clamp

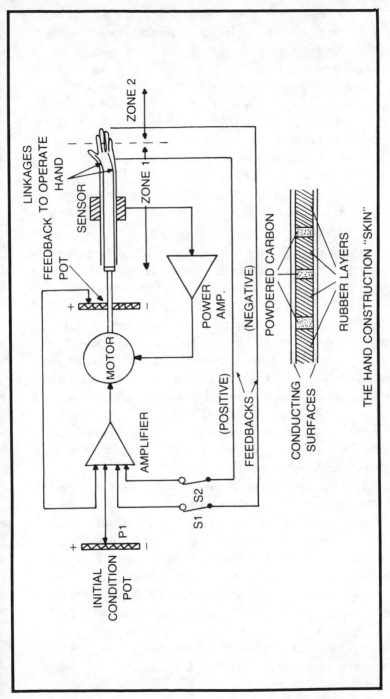

Fig. 7-6. A mechanical hand.

causes a current of opposite polarity. The polarity of the currents determines the direction of rotation of the motor. The pinching (tactile) reaction occurs on contact between the ends of the clamps and the object.

3. In the case of only positive feedback the clamp will develop full force on touching the object. This could lead to deformation and fracture of many brittle objects and also to the useless expenditure of energy. Therefore, regulation of the clamping pressure is introduced as a function of the weight of the object held.

4. There is a switch which enables the operator to use the device as an ordinary servomechanism as he may see fit to do so.

To use this kind of hand on a human body the slave mechanism is executed in the form of a holding device, the feelers (fingers) of which are closed in the pinching position. The design of the drive mechanism of the clamp permits simultaneous movement of all feelers (fingers) of the clamp during grasping and opening. The clamp is controlled by two independent sources of bioelectric signals, one of which controls the grasping, and the other opening. Groups of muscles which are designed to execute the functions of bending and unbending of the wrist and fingers are used as signal sources. A reversing engine (motor) with a special driving mechanism is used as the drive. The developed electronic control junction insures effective control of the power of the actuating drive in case of small muscle tension.

Notice here several important concepts. The use of positive feedback for fast action, counterbalanced by a negative feedback to prevent instability and overreaction. The use of such sensitive sensors able to use the muscle signals to control the mechanism seems almost incredible. And finally, the fact that the hand senses the object and then reacts to it by one of two actions, grasping, which implies full use of all fingers, and pinching, which implies the use of the fingers only, in perhaps a separate sense. An android surely would have to have such hand control. Now, if you add this element to the wrist and arm action previously described you begin to get an idea of the complexity associated with a duplication of the human capability in the android.

IMPORTANCE OF THE SERVOMECHANISM

As we have seen throughout this work so far, the fundamental electric-to-mechanical element in the robot system is a slave mechanism called a servomechanism. It is going to be important to us to examine what this mechanism is and how it works, both from a practical and a theoretical point of view. We will try to keep the theoretical explanations as simple as possible. The big contribution to controlled motion of a servomechanism is that it uses feedback signals which are compared to the input command signals to cause the mechanism to function, that is, to start and stop and move slowly or swiftly as required by the command input operation or signal generated from this operation. We will consider these machines in the very next chapter, and you may want to even expand on what we present with some study of other works currently available on the subject.

Sensors may produce primary information—that is, the command information to a servomechanism, or they may produce the feedback signals to a servomechanism which has as its primary or command input the signal from a prearranged subprogram in the robot's memory. It may be necessary to have many feedback paths for signals. For example the electromechanical hand required many in order that the complex movements of the forearm, wrist, and fingers could be properly executed. We dwell on the duplication of the human arm and hand capability because it will be required if we are to have our robot, or android, do everything we can now vision that we might want it to do for us. Don't we accomplish most everything we do with our forearms, wrists, hands and fingers?

CHAPTER 8
SERVOMECHANISM SYSTEMS

As we begin this chapter, we are surprised that here, on the various corners of our desk, are representatives of the Elves, Gnomes, Trolls, Magicians, and Wizards who have helped us so much in our preparation of our other manuscripts. They have assured me that they are here for a dual purpose. First, to make certain that we do include a little "magic" (mathematics) in our presentation, and second, that we do not make it so difficult and incomprehensible that an ordinary human being will be confused by its inclusion. In fact, the chief troll, with a devilish gleam in his bright eyes, stroked his long beard and threatened to have Klatu, the Imperial Robot, punish me if I stray from these dictums!

So we must proceed with extreme caution, but proceed we must, for the slave machines called servomechanisms are the muscles of the robots. Without them these wonders would be simply giant brains able to do the thinking but never anything else. What a situation that could turn out to be. Imagine needing a little shot of electricity and not being able to plug yourself in!

SERVOMECHANISM BASICS

So what are these machines. They are generally diagramed as shown in Fig. 8-1 and consist of three basic elements. They must have mass (in the load), they must have power (motor) and they must have friction. Realize that anytime you move something having bulk (mass) with an increasing speed (acceleration) or a decreasing speed (deceleration), the presense of mass will make itself known.

The chief troll now admonishes me to state that you should notice your own body (mass) when in your car and you push on the accelerator. As your car speeds up, your body pushes backward; conversely, when you slow down, your body leans forward. These are effects of acceleration and deceleration.

Every machine having mass in its moving parts and in its load will have the same effect. This is related to one of the laws of mechanics which states, "A body in motion tends to stay in motion. A body at rest tends to stay at rest." Keep this wise bit of knowledge in mind as we proceed. Just now, Squeeky, the next to the smallest of the Elves, has informed me that this law is Sir Isaac Newton's first law of motion. If he says it, you'd better believe it! Now it *would* take a gnome to come up with the idea that there are two kinds of friction, static and dynamic. The static friction is what must be overcome when you first try to get something moving. It has to do with the

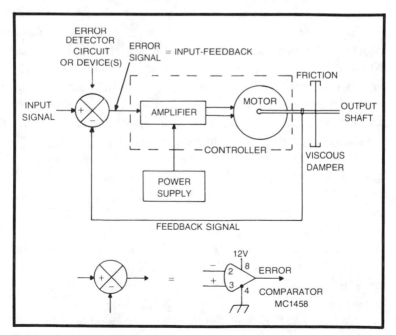

Fig. 8-1. A positioning type servomechanism.

241

molecular attraction between parts of systems in close contact. When you try to push a box along the floor and have a heck of a time getting it moving, you are attempting to overcome static friction. Then when it starts moving, it becomes easier to push. This is dynamic friction. A special type of dynamic friction is called viscous friction (perhaps due to air opposing the motion). Viscous friction is proportional to the speed of movement of the body and finds its home in motor bearings, and lubricated joints which are in motion. The thickness of the lubricant can govern the amount of this viscous friction. If the oil is like syrup you can scarcely move the element, and certainly not too fast. The viscosity of the oil then is great and so its opposition to motion is great. But if the oil is thin there would be less resistance to motion while still getting that all-important lubrication, so the viscous friction now is less.

Power may be from an electric motor. It may be from the hydraulic and pneumatic pistons. It may be from special magnetic arrangements and it might be from the atom, heat, light, or whatever. Did you ever see those four little vanes in a sealed glass bottle turning because of the *weight of sunlight* upon them? Did you know that sunlight has weight or force? Well it does.

But we are interested in the power from electric motors and from hydraulics because that is the stuff which robots are made of. We don't mean to be funny when we add that some robots operate on air. Air is plentiful and we can exhaust it easily without mess or bother. This is something to think about. Suction and pressure also can be used, as it is in many commercial robots for tasks such as picking up or expelling things like papers and tags.

We must realize that we or the robot must have a means of controlling power to whatever kind of motion elements are a part of its body. The Mobot shown in Fig. 8-2 has tank-track mobility and a crane. It was designed for special and difficult tasks in the Atomic Energy Division of Phillips Petroleum Company.

We know that we can control electric power with resistance, no matter what form it may take. Transistors, for example, have their conductivity as a function of the input control voltage. As the control voltage gets smaller the current through them decreases.

We have also been told that the flow of hydraulic fluid through special four-way transfer valves to working pistons is what governs

Fig. 8-2. A remotely controlled Mobot. Courtesy Hughes Aircraft.

the amount of pressure applied to the piston. Of course, pneumatic systems could use an arrangement similiar to hydraulics, but due to possible problems with some types of pneumatic systems we won't refer to it again unless something particular is worth mentioning. But in general for either system, if we have a valve with an electrically controlled opening, and we open the valve a small amount, we will get a small amount of pressure on the piston.

THE BASIC EQUATION OF A POSITIONING SERVO

In science and engineering mathematics the general idea is to write an equation using a set of symbols representing the various quantities. The mathematical arrangement should show the effect of varying each quantity in relation to all the others. The system as a whole can be easily studied. Isn't this a nice idea? If we don't like one solution to the problem we can change the values of one or another or several to come up with the solution that we do like. Wish I could do that with my bankbook! Now this is where the math magicians become important. They know how to manipulate the symbols, in

243

these equations. They can then advise us on which of the various elements we *should* adjust. You see, sometimes it is more important to adjust one than the other.

We will now write an equation which adds forces of a servo-system as:

FORCE due to acceleration + FORCE due to friction + FORCE due to motor = Commanded-Output Force

As you can see this takes into account all the forces we just discussed. Now we need to put this into the symbology of the math magicians. That is really not so difficult, but don't tell them, for the more mysteriously they can camouflage things in symbology the happier they are!

We know that acceleration is change in speed. We know that velocity, or speed, is a change of distance, and we know that if we are to consider forces due to these quantities we must have some means of determining how much force per unit of acceleration or speed is produced. We must also consider the mass present when acceleration is present, and the amount of friction present when speed is present. To indicate that we know all about these proportionality things we will use some letters which group all these factors together. For example, Let M be the constant of proportionality for mass. Let F be the constant of proportionality for speed, and let the accepted letter K stand for the constant of proportionality for the motor force. Now we can write our equation as

$$M_a + F_v + K_x = F_{co}$$

For simplicity we are considering the acceleration, speed, and motor force as acting along a linear distance x. In actuality when moving some robot parts we would be using angles and thus use the symbology θ (Theta) for angles instead of the distance symbol x.

Now if we solve the equation just written, which the magicians tell us is a second order linear differential equation, we can learn how something will react when it has a given mass M, a given friction F, and is moved by a given force K in accord with the speed and position command F_{oc}. Figure 8-3 shows a very elementary analog computer circuit which might be used to solve this equation. Of course digital machines can solve such equations very readily if you have the correct program for them. It is not important what type computer

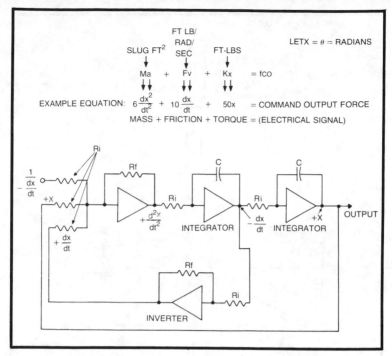

Fig. 8-3. A simplified analog computer for solving second order differential equations.

solves the equation, the important thing is that *it* does it instead of us.

What would actual quantities and movements for all these forces be? In our real-life robot world, the mass might be the robot's arm which is moving upward at some changing speed. That is, it starts at zero speed and reaches a terminal angular speed of movement before it is stopped.

We can determine its *average* speed as ($V_f - V_o/2$). Since its speed is changing it has an acceleration. Friction, both static and dynamic, are always present when something moves and is proportional to the bearing and lubricant used. Motor force *k* is in ft-lb of torque. This is either geared or operates linkages, belts, or pulleys to cause the arm to move. We know that the command output force is really an electrical signal to the motor proportional in amperes or volts to how much torque we want the motor to produce. So we put all these quantities into a computer and let it solve them. When it does it can print out a graph like that shown in Fig. 8-4.

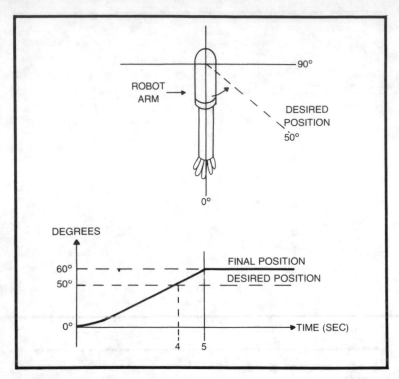

Fig. 8-4. Typical graph for differential equation of arm movement showing overshoot.

Let us analyze what this graph tells us. First we see that the arm starts from a down, or rest, position and comes up to the desired position in a time of four seconds. Then it simply continues going and moves past where it wanted it to stop and comes to rest at a place of its own choosing. What a horrible situation! Why did it do it?

"Aha," you say." The system is simply a motor and gears with mass and friction. There is nothing to cause the arm to stop, as a feedback signal would." And how right you are. You see, we give the arm an up command through an internal or external subsystem. The arm moves up to 50 degrees and we (or the subsystem) remove the command signal F_{co}. But since there *is* mass, inertia, and momentum of the arm, and friction isn't large enough to stop it when the motor stops running, it simply coasts on past its position and finally stops where the friction causes it to stop. That will never do. We need a positive command (not issued by us) for where we or our friendly

246

robot want it to stop. Right? Some persons say that if you use a large enough gear train on the motor, and the arm moves rather slowly, it won't go much past, if at all, the desired position. True. But we want our arm to move as we do, quickly and precisely, don't we. It's unrealistic to have that arm just creep into position.

So now we need to add the element of negative feedback to the system. This can be done by providing a potentiometer at the shoulder joint which will produce a signal opposite in polarity to the inserted F_{co} signal as the arm moves. Then we can use a comparison amplifier and send the result of the comparison through another amplifier to the motor to make it run. Notice that now, with this arrangement, we can get a negative, or opposite, movement of the arm from the signal of the feedback potentiometer if this signal is larger than the F_{co} input. This can happen if the arm moves too far. Let us examine Fig. 8-5. When wiper pin 2 of the physical input

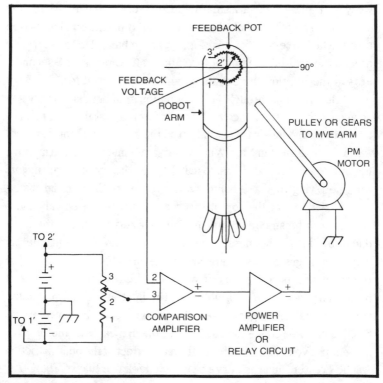

Fig. 8-5. An elementary control system for an arm.

potentiometer is nearer to point 3 than the feedback potentiometer wiper is to its point 3′, the voltage output command will be of positive polarity. When both pointers are the same electrical distance from center, or at center, there will be no voltage output. When the input wiper is nearer point 1 than the feedback potentiometer is near 1′ then the voltage output of the F$_{co}$ will be negative. The comparison amplifier will compare the voltage from the feedback potentiometer to that of the input potentiometer and will produce an output proportional to the *difference* between them, and of a polarity governed by which input is positive and which is negative.

Now let us examine the effect of this on our robot arm movement. We will expect it to stop where it should because if it goes too far, as in Fig. 8-4, then the feedback potentiometer signal will be such that the input will be reversed in polarity, and the motor will reverse and drive the arm down. The only place at which there can be no output from the comparison amplifier is when the arm is where we want it, plus or minus a small dead zone of movement. The graph of this movement now looks like that of Fig. 8-6. The arm has moved up quite fast, overshot its position, and then hunted for a short time around the desired position before finally stopping, after 5 seconds. It is in the position the command signal (F$_{co}$) called for.

"But we went through all the equation stuff but didn't really see an application of it. How come? The answer is that now we need to adjust the constants in that equation so the arm movement is smooth and fast but has no hunting. Whoever saw a homo sapien's arm move up and down, up and down, oscillating as the robot's arm might. Imagine trying to shake hands with a robot whose arm did that.

That ripple at the top of the graph between 2 seconds and 5 seconds can be smoothed out by a proper relationship of the force due to friction F, the mass force M and torque K. It works out this way. For a given mass, and since the arm has certain physical requirements, once we build it we don't like to change its mass. A preliminary value of F, possibly just the built-in friction, and some starting value of K might cause the arm to respond as shown in Fig. 8-6. Now, if we reduce the value of K, which means the motor drive current per volt of input voltage, then the effect of the built-in friction will be greater, and the curve begins to look like that of Fig. 8-7.

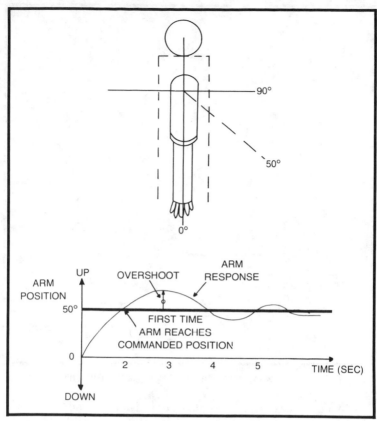

Fig. 8-6. Arm movement with negative feedback.

Notice that it now takes a little longer for the arm to get to the desired position than shown in Fig. 8-6 but it doesn't hunt as much as before. The action is smoother and more life like than before, which is what we want.

In a computer analysis of a differential equation we can adjust the various quantities by simply turning a potentiometer dial representing each quantity, (analog machine) and looking at the curve produced on a scope or graph of paper. We adjust the dials until we get the response we want—which is fast movement, but not too fast. Human speed is desired with no overshoot. Then we can interpret the dial setting into the values of the quantity. Since M and F are normally fixed, we change only K, the gain of the motor controlling amplifier, to get the response we want.

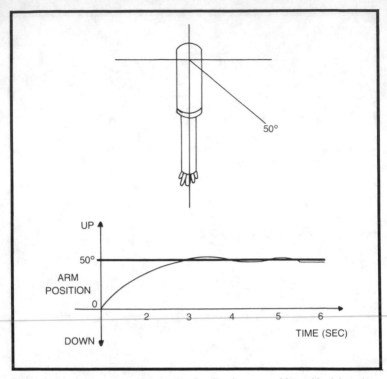

Fig. 8-7. Arm movement with negative feedback approaching critical damping.

If it is necessary to use a motor of a given size and the mass of the arm is fixed, *then* we must adjust the F term to get the response we want. We may not be able to change friction actually, but we can add a rate circuit to the servoamplifier which will give the same effect as increasing the friction, and thus get the kind of response we want by this means. Let us consider the rate circuit next.

RATE CIRCUITS

A rate circuit is one having an output proportional to the rate of change of input voltage. For example, if we have a constantly increasing voltage input, the output would be a constant value of voltage as shown in Fig. 8-8. The form of this rate circuit is, as shown, a capacitor and a resistor. It is like a high-pass filter in a way. The capacitor prevents any dc voltage from appearing across the resistor, and only a *changing* voltage input will produce a current through the resistor, thus a voltage output. The polarity of the voltage is increasing or decreasing. If it is decreasing, after an increase, the capacitor will be discharging instead of charging, and that makes the polarity across the resistance reverse.

Now this circuit is important in servomechanisms. If we consider the symbol of its voltage output to be R (for rate), then we would find—if we make some mathematical manipulations of the symbols in the equations—that this new term will appear as an additive term to the F term discussed previously. The equation will appear as

$$(M) \ (a) + (F+R) \ (V) + (K) = F_{co}$$

and now we see that if we increase the value of the R in the middle term we can get the equivalent (effect) of more friction in the system which, in turn, will smooth out those ripples, or oscillations, in the

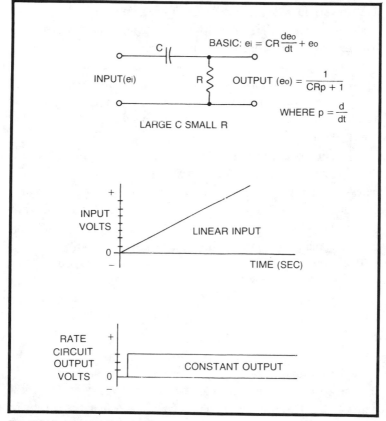

Fig. 8-8. A rate circuit and voltages.

arm movement. This middle term is called the damping term in servo work.

It is possible to get the rate voltage by various means, not just the RC circuit shown. You can use a voltage from the motor (back EMF); you can have a rate generator built into the motor and use its output; or you can use a separate rate generator geared to the output shaft. There are various ways of getting a voltage proportional to speed of movement or rate of change of the controlling voltage or of the output shaft speed. Now let us examine in Fig. 8-9 just how these affect the arm movement when they are properly adjusted, that is, not too much voltage from them, but enough.

In the graph (a) we see that the arm moves quickly to the desired position and then hunts. In the graph (b) we have increased the amount of rate voltage from its amplifier. Now the arm moves up a bit slower, but still pretty fast. And it still hunts a little. In graph (c) we have the ideal adjustment where the arm moves relatively fast but there is exactly enough rate voltage (called error-rate voltage) to produce no hunting. If we have a good motor, not too much mass, and the error rate voltage is properly adjusted, this speed of movement can approximate the human movement under normal circumstances. So the robot's movements become real and lifelike. This is desired.

The error-rate voltage must work in the following manner. If the incoming voltage (the error voltage, which is the difference between the input (Fco) and feedback voltages) has a positive polarity, which is *increasing*, then the rate voltage would want to help the movement caused by it, and so it must produce a voltage which adds to the incoming voltage with the same polarity. When the incoming voltage, due to feedback, begins to reduce in magnitude, but still has the same polarity, then the change in direction of the voltage—the dropping of it or decreasing of it—should cause the rate circuit voltage to change polarity, as the capacitor now discharges. So this should cause the input error voltage to drop very quickly to a zero value.

This change will now cause the drive to the arm to stop before the arm actually reaches the desired position. Then it tends to coast into position. The rate circuit voltage goes to zero and the arm stops. If, by chance, the arm tends to move past the given position, the rate circuit voltage will be based on the feedback voltage polarity. Since this is opposite the input polarity (due to the balanced bridge arrangement) the rate circuit will produce an additive voltage of the same polarity as the feedback voltage. This causes the motor to drive the arm more quickly back to the desired position. We assume

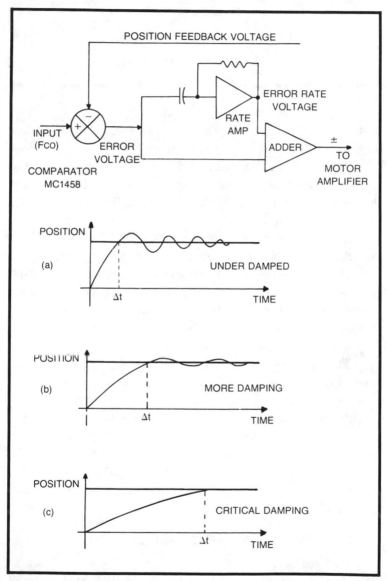

Fig. 8-9. Effects of damping voltages.

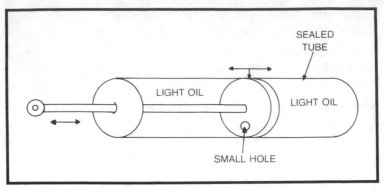

Fig. 8-10. A mechanical damper.

that the capacitor is fairly large and the resistance is a fairly low value to get the rate circuit to operate as it should.

Of course it is possible to use a friction addition in the arm to accomplish the desired result. If you use a tube filled with oil and have in the tube a plunger with a small hole in it, then as the plunger is moved by the arm, there will be a certain drag on the arm, as the oil cannot move very fast through that hole. The oil must be able to move through the hole for the arm to move, of course. The faster the arm tries to move, the more drag is produced, and so the effect is the same as adding rate damping to the servo system moving the arm, the big difference being that the motor must now overcome this additional force and so might have to be somewhat larger. Figure 8-10 illustrates a mechanical damper of this kind.

The adjustment now would be the amount of amplifier gain (K) causing the motor to run. This K factor can be adjusted to give the final touch to the critical damping arrangement, so you won't endlessly experiment trying to get the right size hole in the damper piston. It comes to mind that an airdamper might work here if the piston is sealed tightly, but that is just a thought. Another thing, the thickness (viscosity) of the oil in the cylinder will also have a large effect on the damper operation. Use as light an oil as possible while still obtaining the damping action.

MULTIPLE JOINTS OF ARM-WRIST-FINGERS

We have already seen that the movement of a robot's arm is quite a complex thing to accomplish. There are so many areas of movement: the arm, forearm, wrist, and fingers. Each of these, ideally, should have its own damping and torque application so it won't be hunting as it probes and gropes for objects to do the tasks you want it to. Can you imagine a robot with all these elements undamped. It might shake itself to pieces.

THE MULTIPLE-LOOP SERVO

So far we have described only a single-loop servo; that is, a machine with only one feedback source. This is also called a closed-loop servo because its input depends on what its output is doing. The input consists not only of what we command (or a subprogram commands) but also on what signals come from the feedback or output, since these are compared and the *difference* between them is used as a drive-motor amplifier input voltage.

But as we get our robot to do something (*Anything*, if there is going to be peace in my family. My wife says "I don't want a piece of machinery around here just for looks, or just to talk to! I've got you, dummy!" HMMM!) So we'd better make it do something after all.

We will find as we try to accomplish that goal that we will have not just one feedback loop, but several. And that makes up what is called a multiple-loop servo. That's simple enough isn't it? Now what if there is no feedback, which for some subsystems operations there might not be. This would be an open-loop servo. Reasonable isn't it? Actually we will find that the open-loop concept is important to us not only because some things operate that way but also because we may have to use this idea when we devise a testing scheme for the servo systems of our robots.

So let us examine the block diagram of a multiple-loop servo for an advanced robot. Let's see how it might come into being inside our robot and what purpose and affect all these loops might have. Figure 8-11 shows a Russian concept.

In this diagram, from the Russian report "Robotry Manipuly-atory" we find that the human operator, who may control the robot through wires or radio means, has been included in the block diagram. Imagine writing a transfer function for him. It is interesting that the information sensors not only send back data to the computers through feedback paths—which is what we have been discussing—but also that there is a general overall loop around the whole system

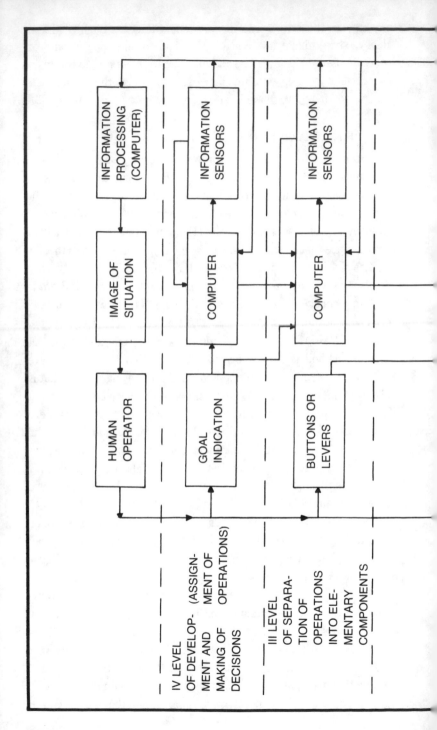

INFORMATION PROCESSING (COMPUTER)

INFORMATION SENSORS

INFORMATION SENSORS

IMAGE OF SITUATION

COMPUTER

COMPUTER

HUMAN OPERATOR

GOAL INDICATION

BUTTONS OR LEVERS

IV LEVEL OF DEVELOP- (ASSIGN- MENT AND MENT OF MAKING OF OPERATIONS) DECISIONS

III LEVEL OF SEPARA- TION OF OPERATIONS INTO ELE- MENTARY COMPONENTS

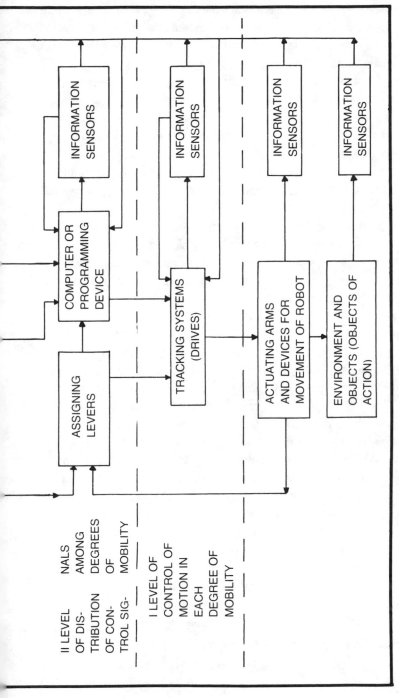

Fig. 8-11. A multiple-loop servo system. (From the Russian report "Robotry Manipulyatory".)

257

so that it becomes a closed-loop system (of very high order) where the input commands are, to some extent, dependent upon what information the information sensors give as to what the robot machine is currently doing.

Look over the block diagram and you will gain some concept of what an advanced robot system might consist of. Of course we must realize that in the completely autonomous robot (that is, one which has no human in its loops) the human operator block shown here would probably be replaced by another computer. For after all, isn't that what the human operator is? Programs or movements would be assigned by this master computer to control everything else, just as the human operator does.

Just in case you are wondering, it is a fact that equations of motion, feedback, rate and such are all written for such block diagrams and then are analyzed by mathematical computers to determine if the system will function with stability or will go wild. Parameters (values of torque, sensitivity, etc.) are then adjusted on the computer so that when the machine is actually built it *should* work properly, and the cost of development will not have been too excessive. It might have been if they just experimented and had to buy more parts and then tried them and then tried again and bought still more parts until they could come up with a workable system.

We will probably start small and economical on our first robot. We will be trying first to get the proper path movement of the machine without having it fall over. Then we might get it to follow a specific path of some kind. Then we might try to have one arm simply raise and lower, then add a pincers grip, the simplest kind, and see if we can have the robot move to various locations and pick up something. (We probably will have to have some radio control over it during this time). Then we might add a second arm similiar to the first. For this we might consider a more elaborate arm which will bend at the elbow as well as raise and lower and grasp with pincers. And so we might continue adding little by little until we have a completely satisfying machine. Beware though. Humans aren't easily satisfied, so this could become a lifetime project!

FEEDBACK WITH A HYDRAULIC SYSTEM

In an ideal robot we might have a hydraulic system with the piston and valve located in the forearm or biceps and we do this because it could be small and have a plunger-piston arrangement which ideally fits into the concept of our own tendons and muscles. It also provides the strength to open and close fingers, or bend forearms with weights on them, or whatever.

A tendon arrangement to close the various joints of the fingers could by very desirable if we are to pick up objects. You might want to consider counterbalancing of your robot's arms if they are heavy and hard to move.

THE NATURAL FREQUENCY OF A SERVOMECHANISM

Everything in nature has a natural frequency. That may be a broad statement but it could be seen to be true if only the extent of our technical observations were able to observe this kind of oscillation. A servomechanism is no exception. It also has a natural frequency which, for a simple sytem, can be expressed in a mathematical way as

$$\omega_n = \sqrt{\frac{K}{M}} \quad \text{or} \quad \sqrt{\frac{K}{J}}$$

where

ω_n = Natural Frequency
M = mass
J = inertia
K = force

The result is expressed in radians/second. Notice what this tells us. If our mass is large (the arms heavy for example) in proportion to the amount of driving torque we have for that arm, the frequency of oscillation will be slow. If the mass is light, the frequency will be fast. We have introduced a term J here which stands for the mass (inertia) in a rotating system. It is used exactly the same as mass M is in a linear system. So we generalize by showing both terms. In the rotary system we find all the inertia, that of gears and shaft and load, lumped into this J term.

The arrangement of terms in the natural-frequency equation gives us a clue as to which to adjust if we are testing a system. If we have a high frequency of oscillation we can reduce the K term, probably by reducing the drive to the motor amplifier. On the other hand if the oscillation is very slow, we probably don't have enough power so we might have to increase it. This latter condition assumes of coure that we cannot change the mass of the arm as for example from steel or iron to aluminum. Perhaps we already have the mass in aluminum. We can't then change that to a lighter material.

Note however that counterbalancing a heavy arm might give the effect of reducing the mass, as it would then be much easier to move. But since the effect of the masses are present when accelerating the movement, adding a counterbalance also means more mass. So the arm might move still more slowly as far as natural frequency is concerned.

Thus we find now that the hunting of the servo-output element, such as the arm of our robot, will have a definite frequency of movement. This leads us to look at one solution to the differential equation representing this type servosystem. The solution to the addition of forces is not as simple as just adding them. When we get a solution by classical means, not the transfer function approach, we have an expression (the solution) which looks like

$$X(t) = e^{-at} (\cos \omega_n t + \sin \omega_n t) + A$$

The $X(t)$ represents the instantaneous position of the arm in, say, centimeters or degrees. The exponential term (e^{-at}) shows that the size of the oscillation will decay in time by an amount represented by the a factor value. This a factor is the term which we know to be $(F + R)$ back there in our original equation. ($\omega_n t$) is the oscillating frequency ($\omega_n = 2\pi f = \sqrt{(K/J)}$ in radians/second. And A is the so called steady-state error, which is the amount of angle or distance that the arm lacks to reach the *exact* position we commanded it to reach. A graph of this solution looks like that shown in Fig. 8-12.

Notice that we could expect with this under-damped system that the arm would come to its desired position quickly, then hunt for just a cycle or two, then stop a little off its final desired position. The final error might be so small that we could ignore it, but for high precision and high resolution robots, such as are used in industry, that error might have to be so small in some cases that it would be almost nonexistent.

Refer again to the natural frequency expression $\sqrt{(K/J)}$. Look at the numerator under the radical sign. In this very simple expression we have not made any provision for the effect of the friction term F nor the rate term R. Let us remedy that situation right now,

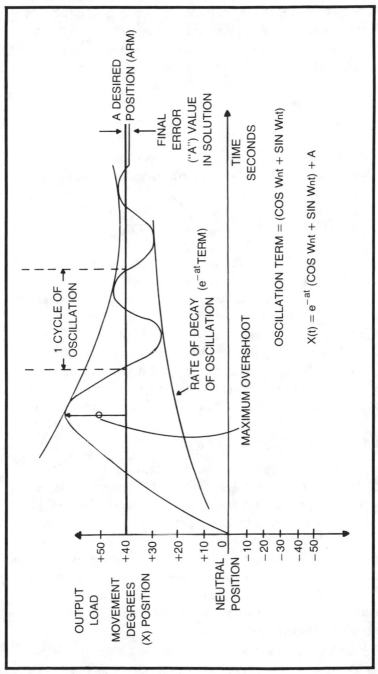

Fig. 8-12. A second order differential equation for an underdamped case.

OUTPUT
LOAD
MOVEMENT
DEGREES
(X) POSITION

+50
+40
+30
+20
+10
NEUTRAL 0
POSITION
-10
-20
-30
-40
-50

A DESIRED
POSITION (ARM)

FINAL
ERROR
("A") VALUE
IN SOLUTION

TIME
SECONDS

1 CYCLE OF
OSCILLATION

RATE OF DECAY (e^{-at} TERM)

MAXIMUM OVERSHOOT

OSCILLATION TERM = (COS $W_n t$ + SIN $W_n t$)

$X(t) = e^{-at}$ (COS $W_n t$ + SIN $W_n t$) + A

261

for we know that if we add friction or rate damping this will affect the frequency, and time, of oscillation. We have even gone so far as to state that we could eliminate the oscillation altogether if we have the right amount of F or R, or the combination of both in the system. By the way, these are called retarding forces because they oppose motion when used in the damping concept.

The oscillation in such a system as this will approach zero as the relationship

$$F^2/4J^2 \text{ approaches in value } K/J$$

We elaborate on this to say that $(F+R)^2$ will be used instead of simple F). We now look at the quantity under the radical and see that it now turns out to be

$$\omega_n = \text{NATURAL FREQUENCY} = \sqrt{\frac{(F+R)^2}{4J^2} - \frac{K}{J}}$$

As you know we can adjust the value of R so that there would be no question that the first term of this equation would numerically equal to the second term. Then, as is shown by the fact that we have zero under the radical, the oscillation will be zero. The servosystem will not hunt. The arm will move to the desired position with no over-shoot and no oscillation. That is what we want in our robot system.

So what? We knew that all the time. What have we learned that we didn't already know? We answer that with some math magic. We won't show (no good magician does) just how we manipulate the previous expression into

$$(F + R) = 2 \sqrt{KM} \text{ or } 2 \sqrt{KJ}$$

but we assure you it is good and proper magic, and this is the important additional knowledge. You see, this mathematically tells us how much damping is required for any combination of force K and mass M or J in order to achieve the critical-damping condition of no oscillation. The backforce, or damping force, exerted must be such that by adjustment we get a value in 1lb-ft/second of friction, or friction plus rate, or rate voltage, equal to the value on the right side of the expression. This helps in design and we design the elements of the rate circuits to be of such gains and voltage outputs that we can easily get the required value.

If we study hard and learn a little of what others have learned about all this, we can design a robot on paper before we start buying parts and putting them together. That doesn't mean that the paper design will work perfectly the first time we assemble it. Not by a long shot. But it will give us a good starting point if we really want to get in to the serious design of robots and then construct them. Researchers love college campuses and company research labs. You'll find them there.

DIMENSIONS

One final word about something called dimensions, and it isn't a rock group. It has to do with the quantities represented by the letters in our equations. The K for rotary (motor) systems is expressed in output torque per unit angle of error: ft-lb/radian or dyne-cm/radian. The friction torque F is dyne-cm/radian/second and the mass in a rotary system is expressed as gm-cm². In the British system J is expressed in slug-ft², R and F are in ft-lb/radian/sec and K is ft-lb/radian. If you are mathematically inclined and decide to work with these, be sure to keep your dimensions consistent.

ANOTHER METHOD OF OBTAINING A RATE VOLTAGE

If we obtain a rate voltage from an output shaft through a rate generator we mathematically find that this has exactly the same effect as an input-rate voltage. The two methods of obtaining rate voltages are shown in Fig. 8-13.

To get the output-rate voltage we use a small rate generator geared to the output shaft. If we gear it to the arm shaft of a robot we must realize that the arm shaft will probably move slowly. So to get a good voltage output we will want to use a step-up gearing ratio to the generator. This will make the armature turn fast while the arm moves relatively slowly. This way we get a good voltage output which we can adjust with a potentiometer.

Notice that the rate voltage is applied in series with the input command signal, and in this manner will add to or subtract from the input voltage, just as two batteries add or subtract voltages when in series.

The small tachometer generators can be obtained from the same manufacturers who make small servomotors. A check of their catalogs will show what is available for this type system. Again we

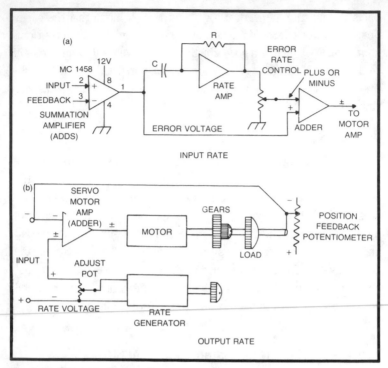

Fig. 8-13. Sources of rate voltage.

emphasize that it doesn't make any difference whether you get the rate voltage from the input or the output. As shown, both will give effective damping in a robot servo system.

SUMMARY

The servo mechanism is a vital part of any mechanism which is going to combine the electrical signals from somewhere to produce a mechanical output. We have examined only the most elementary positioning servos, which incidentally, will do the job in most hobby robots and even in some industrial types. As we have also seen, however, the more elaborate robots will have many many interacting servo systems. Then the mathematics and analysis become more complex. We suggest that when you build your robot you should breadboard controls, servo circuits, and motors and get everything working properly. Only then consider installation in the robot body. We hope and trust that then your robot will work as well as threepio.

Chapter 9
Commerical Robots

Our most spectacular robots find themselves being blasted into space for long journeys to distant planets. These represent the ultimate in technology. They carry sensors for measuring everything from temperature to organic soil analysis. And their cost is in proportion to their capabilities. But the spin off from these wonders soon filters down to less-exotic applications. We'll look at some of these in this chapter.

MEDICAL APPLICATIONS

A story is told by the robot master of Quasar Industries about a robot application in a medical facility. It seems there was a child undergoing treatment for a trauma wherein he had lost his ability, so it seemed, to speak. Every kind of treatment had been used and still the child just stared into space and wouldn't communicate, even with his heartbroken parents.

It was suggested, and the idea accepted for a limited application, that Quasar provide a robot companion for the child. This robot was to be dressed in clothes similar to that which the child would find on his favorite toys or on his previous friends. The reasoning was that if the child would not communicate with adults he *might* communicate with an inanimate object such as the robot resembling in dress, at least, his toys and friends.

When at first the robot was placed in the child's room the child gave no indication that he had seen it, was aware of it, or that it in any way changed his world of silence and fear. But then the robot would speak to him now and than saying "I am a robot. What are you?" And the voice was friendly and not exactly human, so a response was expected. The child didn't seem to notice. Then the robot was caused to move around the room, slowly, and carefully, but the child still paid no attention to it.

After some time, when the staff began to consider that this, too, was a failure, the child was put into his bed and the adults left the room, and the robot was made to go over to the bedside. "I can speak, you can't", he said, moving an arm. The child didn't respond. Then the robot began to back away slowly. "You're a dummy; you're a dummy, you can't speak. I can speak and you can't. You're a dummy," the robot said excitedly.

And the child looked at it. "I am *not* a dummy," he said, to the utter delight of everyone concerned.

Well, we don't know how other kinds of treatment or time might have affected this child, but we do know, according to the story, that he recovered completely and that his grateful parents felt that their prayers had been answered through the use and appearance of a friendly robot. This one was controlled by an observer and spoken through by not only therapists but also the robot master who gave it a personality such that the child felt some emotion and responded verbally to it.

There can be no question but that in future years we will see more and more applications in the medical field of the use of robots. These will monitor life support systems, furnish patients with physical care, entertainment, and companionship, and be as essential to that field as are the other medical machines now currently in use. Some robots, but not in the forms which we usually think of as a robot having, are currently in use in medical facilities. These machines have the big factor in common with other types of robots. That is, they do something, and what they do is dependent on what the output, or patient, is doing.

Consider anesthesia. When a pateint is given a drug to keep him unconscious so that an operation can be performed, brain monitoring devices can pick up brain waves and use these as a control factor for determining the amount of machine-administered drug put into his veins. The brain activity becoming quiet and smooth is an indication that he is under and ready for operative care. If the brain activity gets "loud" (as the machine sees it) this indicates he needs more drug and so the machine causes a little more drug to be pushed into his veins via a hypodermic injection. Some tests have shown this to be a way to administer anesthesia.

Of course the robot medical machines can monitor may other life functions including heartbeat, breathing, and temperature, as well as brain waves. It can assess all of these to determine what and how much stimulant or depressant should be injected. The machine doesn't get tired and its attention never strays away. Its power comes from a dual system, one an emergency system so that it is never without power regardless of what may happen in the hospital. The exception of course being total destruction or a major disaster.

The essence of these robots would be a "brain" such as we already discussed. They might be motorized so they can do the jobs required in this application. They require sensors which can furnish input information from the many sources medical experts determine are necessary. A mistake by the machine or failure of an internal part might cause a patient to die or to be rendered incapacitated in some manner. That has to be overcome by a redundancy of systems. The machines used in those locations nowadays required that the reliability be virtually perfect. One cannot be perfect, remember. Even humans have their failure rate, and it just might be much greater than a machine's.

What about redundancy, the duplication of systems? Is this feasible and probable? The answer is yes. This is already being done in spacecraft, aircraft, in atomic energy monitoring devices, etc. There is a branch of mathematics which enables designers to determine just how much reliability is needed. This leads to the kinds of systems needed to insure that reliability will be met. Expensive? Yes. But the answers are there for the calculating.

INDUSTRIAL APPLICATIONS

Many industrial robots are now working, and more are in the development stage in countries all over the world. The U.S., Russia, and Japan lead the field, and the results over the coming years will

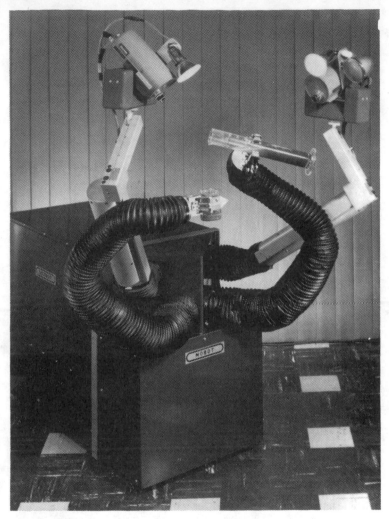

Fig. 9-1. A Mobot handling chemical transfer. (Courtesy Hughes Aircraft)

lead to more automation in plants making our everyday household items and other devices for our pleasure and comfort. It could be that in time we will be able to purchase automobiles which do not have to be recalled because of a missing bolt or screw. There's the story told about one owner who had an annoying rattle in his car window mechanism. So he had it replaced, only to find in the process, a note written by some bored factory worker which said, "Took you a while to find it didn't it?"

268

Human errors and frustrations will be eliminated and processes will become more and more perfect with the diligence and perfection that only machines can produce over long periods of operating time. And each machine will have sufficient error detection devices and failure alarms so that it will be almost incapable of performing any part of its operation less than perfect. As of this writing, Westinghouse has just been awarded a quarter million dollar grant by the National Science Foundation to study the feasibility of applying the programmable type assembly systems to batch manufacturing. This is because it has been estimated that 75 percent of our total outlay for manufactured parts is accomplished by batch manufacturing methods where the lot size is 50 units or less. It is believed that complex tasks will still be performed by people, but many repetitive, boring tasks, and tasks performed in unpleasant atmospheres or dangerous conditions will be performed by machines. Figure 9-1 shows how a Hughes Aircraft Mobot might be used to handle a dangerous liquid transfer. Figure 9-2 shows the control operator. This might be a job classification for persons in the future.

Fig. 9-2. A control panel for the Mobot. Courtesy Hughes Aircraft.

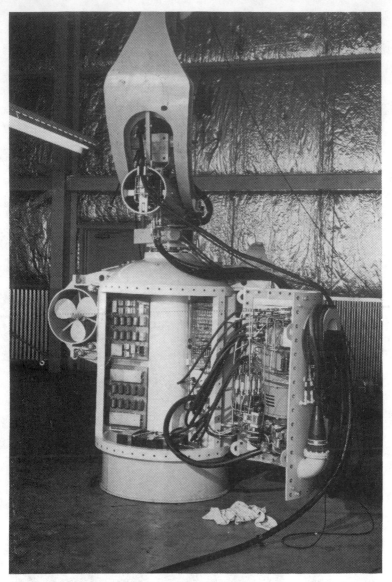

Fig. 9-3. The Welmo Mobot for underwater exploration. Courtesy Hughes Aircraft.

UNDERWATER APPLICATIONS

One of the largest areas where robots will become mandatory, resulting in a great demand for complex robots is in the exploration and development of under ocean operations. These robots at present do not exactly match our concept of what a robot should look like. It is true they have external controls—which we shall explore in a later chapter—but they function as robots in that they *do* jobs for us under water at our command. One such robot is shown in Fig. 9-3.

Notice the small propeller. There are actually two of them to permit movement underwater to specific locations or over specific underwater terrain. The complexity of the electronics is such that good job opportunities will exist here for persons trained in instrumentation electronics.

Figure 9-4 shows the Welmo robot being tested in a large tank. The control console appears in Fig. 9-5.

It is interesting to note that the word Mobot was derived from the words mobile robot by the manufacturers of these machines.

Now let's look at Fig. 9-6, another industrial robot, which was specifically designed to operate various valves fitted on an oil wellhead located underwater. There is a device called a Christmas tree fitted to these wellheads. These consist of BOP's (blow out preven-

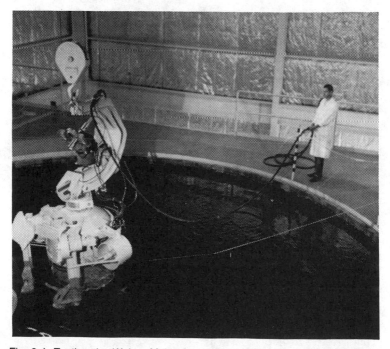

Fig. 9-4. Testing the Welmo Mobot in a large water tank. Courtesy Hughes Aircraft.

Fig. 9-5. The control console for the Welmo Mobot. Courtesy Hughes aircraft.

ters), valves, couplings, etc. There is also a track around the wellhead to which the Mobot can be attached. With its television eyes the Mobot can be maneuvered to any valve or bolt, and then it can move as desired to manipulate these elements.

The Mobot is equipped with a television camera, two mercury vapor lamps, a hydraphone (to listen), a sonar transducer (sensor) a gyrocompass to give it a sense of direction, a depth indicator, and other instruments to show things like hydraulic pressure and flow. The Mobot's functions are driven from a hydraulic pump having motorized controls to adjust the oil flow and pressure. The TV focus and the flow and pressure controls are electrically powered. All other operations are hydraulically powered.

This Mobot, developed for underseas work by Hughes, is large. Its size can be judged from the man beside it. It has two manipulating arms which are six feet long, and each one has a

Fig. 9-6. Unumo, a universal underwater Mobot. Courtesy Hughes Aircraft.

shoulder, elbow, and wrist joint. Each joint has full hemispherical motion (180 degrees), which means 90 degrees in either direction from neutral. The wrist can rotate in both directions and can extend about 3 inches. The hands are removable, as there are several different hands designed for this Mobot. Of course these manipulating arms are always in good view of the TV camera.

The two bottom arms are called clamp arms and are similar to the manipulating arms except that they are shorter in length and larger in diameter and do not have a wrist. These arms are extended to hold the Mobot steady while it does work with its manipulating arms. But they can also be useful for picking up objects from the ocean floor and for handling heavier objects than the manipulating arms can handle.

Unumo is operated from electrical and hydraulic power supplies. It can maneuver angularly or linearly in the water by means of two propellers, one on each side of the unit. On top of the Mobot is a propeller which provides right or left translatory motion, or bending motions—fore and aft.

The underwater Mobots are suspended from the mother ship by a wire rope which has the electric control cable attached to it. They have built-in bouyancy tanks which will cause the Mobot to rise within 20 feet of the surface in case of emergency. The arms are powered by a three-phase gear-head motor, meaning that the gears are built-in to the motor itself. These are controlled from the console using level switches similar to those used for radio control of model airplanes. One switch operates one entire joint, such as an elbow left-right and up-down motions as well as a combination of these motions simultaneously. This is very much like the radio controlled model aircraft in which you control rudder and elevator simultaneously. But also, on the Mobot you will find some levers which have only two directions of motion. These operate wrist extend and retract motions and wrist rotate left or right actions.

In the original version of this Mobot, two rotary switches were used, and each had 64 contacts brushed by a rotating wiper. There was one on the console and one in the Mobot. They were exactly synchronized at a speed of 1800 rpm. If you wanted to send a signal to switch contact 4, you pointed the console switch at its contact position 4. It would pick off a voltage there which *at that instant* would be sent through its pointer to the pointer of the underwater switch which would be on its contact 4 also. This voltage then, after averaging, would cause the Mobot to do something.

Notice that with 64 contact positions you could control 64 functions in the Mobot. The only problem with this system is maintaining exact synchronization of the two motor-driven pointers; otherwise a signal which should go to contact 4 of the Mobot might go to contact 5, and this could cause lots of trouble. The system operates so fast, 1800 rpm, that all 64 functions seem to be simultaneous and instantaneous in operation.

The voltage is picked off a contact of the Mobot switch and goes to a driver-card circuit which has a holding capacitor. This permits a build up of the voltage across its terminals if the voltage persists. It would persist if the signal was a command. If it was not a command but static or impulse noise, then it would not build up on the capacitors. Thus the system is made somewhat immune to noise impulses.

The capacitor voltage then is applied to two Darlington circuits, one for positive inputs and one for negative inputs. These, in turn, operate relays which control the voltages to the function motors. Think about this system. You could probably build one like it if you can solve the synchronization problem.

It is probable that pulse techniques would be useful in some applications to control many functions. But to resolve a train of 64 pulses might be somewhat complex in circuitry. Of course computers do this. Model airplane radio control systems use eight pulses in a train to control the required functions such as rudder, elevator, flaps, ailerons, motor speed control and bomb drops or camera controls. A model airplane radio control system such as is available on the market might be used to give you control over your first surface robot. There would be no wires attached.

SPACE APPLICATIONS

These, of course have to use an autonomous system to cause operations to be performed. Preprogramming of subsystems, which are computer controlled and which can be activated by earth commands, form the primary source of intelligence in these systems. There is reaction to information obtained from sensors such as gyroscopes, inertial systems, accelerometers, star trackers, and sun and planet followers which keep the units under their control performing as they should. In the future there will be more self-determining systems in space explorers. We will not probe into depth as this information is available from NASA.

SECURITY APPLICATIONS

Ever since such movies as "The Day the Earth Stood Still," we homo sapiens have wistfully wished that we could have a guard and protector like the giant robot Klatu, who saved his master from an earthly death. In general we imagine some giant machine, human-like in appearance, having strength beyond belief and power to paralyze our enemies with light rays, sound waves, or unknown forces which it alone can control. It would be almost as if we then had the capability shown in Fig. 9-7 of the Hardiman human attachments built by General Electric to increase our human stength.

In the equipment shown, the arm and leg movements are perfectly duplicated, but with the strength of a giant. Under human control and equilibrium the machine parts will walk and give hand and arm movements which are exact duplications of the human ones. This is the thing we envision in the ultimate robot, but without the human brain in the system. When the human in this figure can be replaced by a computer and sensors the autonomous robot will become a reality. That day may not be far in the future.

The Quasar Security Robot

As reported by UPI there is a robot built by Quasar Industries called Century I. It said that he is 7-feet tall, weighs some 650 lbs, and was specifically designed to be a security guard. He is said to be bulletproof and equipped with all kinds of retaliatory mechanisms and devices which should render an invader helpless. It is said that he can detect movement, body heat, or any noise, and then will lock in on that phenomenon and go after its source. When it gets within 8 feet of the target it will verbally instruct the intruder to stop dead in his tracks, and if the person doesn't, then it may employ such devices as a very highpowered, high frequency audio to cause extreme pain to the inner ear; it may blind the intruder with an extremely bright flashing light; it may shoot an electronic dart which delivers a high

276

Fig. 9-7. Hardiman I, under human control and equilibrium, provides great strength. Courtesy General Electric.

voltage shock, or it may deliver a volume of laughing gas in a squirt.

In Fig. 9-8 you will see another kind of robot, a Hughes Mobot which, without the crane, might be used as a security guard. Can you imagine the terror in an intruder's mind if he saw this monster coming at him at 20 to 30 mph with all arms waving and the thing shouting at him and sending out other alarms such as unearthly screams and siren sounds. Golly!

In the space now occupied by the crane and hoist mechanism would be all the sensing and firing devices needed to make it a complete walking equivalent of an army. Of course it would be internally programmed to function automatically unless you send it the key signal immobilizing it. What a big secret *that* signal would be!

"Someone might blow it up," you say. True. Anything is possible of course. But in doing this the intruder also gives an alarm and so the objective of his intrusion might be foiled. Since the machine has such as fine mobile system (tank tracks) it probably could be armored against all but a gigantic explosion, just as other military tanks are. But also, how many burglars are going to that length to defeat such a security system, and if the robot senses explosives he might detonate them in the intruder's hands!

Abilities of a Security Robot

If we let our minds dwell on the subject of security robots for a little while we come up with some enlightening concepts. First, a robot might receive information via electronic means from seismographic sources placed in the ground at strategic intervals all around the guarded area. These can be made so sensitive that even a cat walking can be easily detected. The robot's electronics knows from which sensor the signal came and would hurry in that direction. At the same time the robot sends out a silent alarm to law enforcement agencies. Night vision sensors are better than human eyes for movement and object detection, and the robot's memory knows if the scene it sees as it looks around has any added elements or not, even if there is no movement.

Now, the fact that the robot is patrolling the grounds means that he will be at the scene of the intruders activity before law enforcement personnel can arrive, and so this would be a good deterrent. Of course there can be other fixed sensors around the area to give the robot information about movement or presences on the grounds.

Fig. 9-8. A Hughes Mobot on tank tracks. Courtesy Hughes Aircraft.

We wondered about the advantage, if any, of using this robot in place of police dogs as patrolling devices. The cost is higher, the system more complex, and the maintenance more demanding. But as we gave this consideration it seems that with all the responses possible to fixed sensors they are an advantage. And there is immunity to bullets and poison, which an animal might fall prey to. The robot might have some advantage over the animal in spite of the disadvantages.

There is another aspect of the patrolling robot which warrants mentioning. With modern technology it is possible to give it a random patrol movement. Since this means the robot doesn't know where he is going next, or how far, or how quickly, how could an intruder possibly know what that machine is going to do next? With human watchmen it is known that they make the rounds on a time schedule. Even though this may be varied it has a pattern over some weeks or months which can be determined by diligent observation. With a random path generator in the robot the course of patrol may never be the same and just as unpredictable as random numbers are in the mathematical world.

LEARNING PROCESSES

We have mentioned without much explanation that a machine can "learn" from a human thru signals that put it through its motions a time or two. Then the robot's memory will enable it to duplicate that series of movements time after time without deviation or failure. Let us think about the learning process a moment here and see what actually is done to make this process a physical reality.

First of all we might think of a learned movement as a command input signal into the robot's movement servosystem. This means that command signal will have polarity, duration, and amplitude, or will consist of some orderly arrangement of pulse or tone signals. So what we must do when we "teach" a robot is to provide it with a preset program of signals, probably repeated over and over again, as all movements in the cycle of operations are accomplished. We will be sure that the robot's movements do what we command because the feedback systems of the servo will insure an exact response to the commanded input.

Now suppose we use a simple magnetic tape and attach it to the feedback potentiometer of a shoulder axle. If we cause the arm to move to certain positions by generating command signals, then the feedback signals recorded will be negative reproductions of the commanded signals. An inverter can convert them into positive signals so they can be used as a command input when the human doesn't control the machine. The machine has learned a series of operations which will be repeated as long as the tape memory is caused to run in an endless cycle.

You will no doubt think of other means of recording the command voltages. Perhaps the best way is to send them into a small computer memory bank where the voltages will be permanently recorded and can be sampled at appropriate times in a process of robot movement. There also might be a film strip which has light and dark areas formed on it as a light varies its intensity during learning when the human operator causes the robots movements and actions. When played back using a light and photocell pickup the original command signals are again generated.

This is the stuff of which robot memories are made. They can be mechanical, electrical, or electronic. The only criterion is that they be able to cause some variation in something of a permanent or semi-permanent nature which can then be used to create the input signals to the robot's machinery.

AUTOMOBILE APPLICATIONS

Automobile robots will not have the semihuman form we like to associate with these machines, but they are robots just the same. In the future you will have safety collision radars as a part of the automobile sytem which will require you to automatically maintain a given distance to the car ahead, and will rapidly apply brakes and ease up on the gas when the distance is less than required for safe driving. Warning systems which will give information on side approaching automobiles or of other dangers will be incorporated into the automobiles of the future.

Most everyone has heard or read of the pilotless automobile. This will follow a prescribed path along a roadway according to what you set into it by means of a route-key system. Its computer brain will do all the work for you. If you are going to work, just climb in, set the controls, and the auto takes off, safely maintains its place in traffic, and finally reaches the destination. It glides into its prescribed parking place at your place of business, or lets you out and drives on to its own parking place, to return for you when you energize it through a proper signal later.

You can read, study, or watch TV during your riding time, or as on trains, enjoy a friendly hand of bridge with fellow commuters. To go out of town you set a different program on the auto computer. You are steered out onto the highway of your choice where you can, if you desire, take over in response to a suitable warning or alarm, maybe the friendly robot voice saying "It's your turn now."

It may be too expensive in the immediate future to have these path controls in cities and on state highways, but much experimentation is now going on with controlled transportation. Of course you already have many robots acting in your car. There are the air conditioning and temperature controls, lights which come on and off and dim and brighten according to the cars approaching, fuel injection systems operated by small computers designed to get the maximum fuel efficiency, and computerized trip-o-meters which give you everything from distance remaining to average speed and engine rpm. Automatic gear shifting already takes place and we do not even think about this. In aircraft the number of robot operations is even more impressive, and you wouldn't believe what happens in a modern ship nowadays. So we are already living with robots although we don't recognize the forms. We simply think of them as automatic systems.

MOON EXPLORATION APPLICATIONS

One of the Russian space robots sent from earth, landed, drilled for a sample of moondust in a rocky area, then took off and flew back to earth and landed itself in a blizzard where it was supposed to land. The magnitude of this accomplishment should not be underestimated, for the robot was almost entirely under its own control in this process. This was the Russian Luna experiment 16. When we say almost we mean that some path information was provided by humans, which the robot was incapable of calculating on its own. But just think of what this means in terms of the possible actions of unsupervised robots of our own future. We have explored Mars with a similiar machine, haven't we? It didn't return to Earth, but then we got all the information. Yes, we will see more planets, maybe even stars, explored by robots. Of course a star explorer would leave here during one generation and return, perhaps, several generations later.

AGENCIES STUDYING ROBOTS

Manufacturing corporations such as Hughes, Westinghouse, General Electric (who have all contributed to this work and we thank them sincerely for their help), AMF, Hitachi, and the telephone companies are involved in the study, design, and development of robot mechanisms, both from the standpoint of solving particular problems, and in trying to find out what *can* be done with these machines. There are also many universities and even government agencies who have grants to analyze and study and develop these systems. They study the mathematics of design and the algorithms of operation then construct and operate robots to prove their abilities.

Such well known places as MIT, Stanford, University of Aston (England), Oxford, and NASA are among the leading participants. The development is proceeding in the U.S., Russia, France, England, and Japan. Japan has even made a whole factory of robot machines (no humans except supervisors) to test out the future capabilities of automated mechanisms. This again means that we will have more and more robots around us as the demand increases and as the fear of using robots declines. This fear is not just of the robot's actions in itself, but also in the effect on our economy (jobs) which will undoubtedly result.

One has to remember, however, that as we went from covered wagons to the automobile, jobs increased because of the new requirements. Instead of a few blacksmiths we began to need many technicians and design engineers. And so it will be with this new robot field. More jobs will come about, and although many will be different from some jobs we do now, they will be challenging, exciting, and fun. It makes one want to get into this field early with design ideas incorporated in one's own robots. We should learn as much as possible in preparation for what is to come.

WAR APPLICATIONS

Yes we have these. They are the guided missiles, cruise missiles, underwater attack devices, space hunter-killer satellites, and even ground roving robots for battlefields. These can monitor and deliver gear into a battle field or other suspect territory. War robots are highly developed as of this writing. Countries give them a high priority for research, development, and construction. They use earth phenomena such as magnetic field, astronomy, electric field, gravity, and physical identity (topography) to give them knowledge of where they are and how to get where they have been told to go. And many are immune to known countermeasures. Once programmed and launched, they "take over" and that is it.

What about robots fighting robots? It's true, yes. There are counter robots to fight robots. And well do they perform. We have witnessed the interception of rocket robots in flight, high in the atmosphere, by other rocket robots which have been commanded to "Kill the Invader!"

There are other hunter-killer satellites robots currently in existence. We could have a situation where battles would be fought by machines instead of by people. It could happen above the Earth, in space, or under the sea, or on land. Right now, for example, the British are using remotely controlled and automatic robot machines in Northern Ireland to investigate sites where it is possible that bombs have been planted. These bombs, if present, are detected and then exploded by the robot without harm to humans.

But the development and analysis of war robots will also have its good side. The techniques will be adapted to civilian use as time goes on. There is much which cannot be revealed at present, of course, due to national security, but as people become more peaceful (is it possible?) and means other than war are devised and accepted to settle arguments and differences, then we will find these robots revealed in the manufacture of pleasant and efficient devices to make our own lives better.

Probably the most popular robot will be the housewife helper wherein you sit down at the dinner table on a special chair. It will quickly and efficiently weigh you, obtain a blood sample (painlessly), and send it for an instantaneous analysis to the home computer which also has kept track of your age, general condition, height, and activity level. The chair will also take your temperature (by a new means). Finally, in just a minute or two, it will automatically set before you a specially prepared meal to keep you lean, healthy, and alert to prolong your life to the maximum extent possible. Dishes will vanish when you have finished eating and the clean up will be automatic. Now if you happen to have a "bug" which the blood sample will show the computer, it will quickly analyze this and give you a "shot" of what's proper and needed and in the right place without pain. Isn't it nice to dream.

Well, such are the industrial robots and some fanciful thinking. Perhaps there is more truth here than we imagine or know about. We do know that what industry is doing and has and will do is no flight of fancy. Industrial robots are real. But since they are a little beyond our own capability to construct and learn about, let's get some construction tips about things we can fabricate—in the next chapter.

CHAPTER 10
TIPS ON
CONSTRUCTION OF HOBBY ROBOTS

If at this point you have some slight desire to build a small scale robot to experiment with and have the fun and challenge of this new hobby, then join us in this fascinating chapter. It is a real challenge because development is unlimited. Building a robot offers a real reward in that you have something friends and neighbors will look at in amazement and then at you in admiration and with respect as the one who conceived and built and can set into operation this precursor to the android—a real robot.

WHERE TO START

We won't give you step-by-step construction guidelines in this chapter, but rather will try to provide you with some knowledge necessary to build and construct such a machine. Some of these constructional details, as you would want to fashion them, might be very different from what someone else might want to have in his machine. And that's really part of the fun of this hobby, taking raw knowledge, as it were, and building and elaborating upon it with your own concepts and ideas to come up with something new and amazing. Now a word to those who might be more advanced than others. There is much literature available on the subject of robots, and this ranges from the simple to the very abstract and complex. If your ability and education warrant, we recommend literature from college libraries and research and real laboratory experiments. Much of this combines computers with automation in an effort to develop new techniques and uses for robot applications. The field is large and the rewards can be great.

We will start from a very basic state in this chapter. Now that we have some ideas already, fabrication itself becomes the challenge. You might have an idea but just how to get it into a physical and operational form can be as challenging as design, mathematical proofs, and blue prints. It may require special cams, levers, and gears which must themselves be dreamed up, designed, built, and tested.

Let's begin here with a simple steering and power unit which can use the smallest of toy motors or larger motors for a useful end product rather than just a demonstration toy. From this we proceed to consider some additions and expansions which might be used to make a robot the well behaved machine that it ought to be!

BASIC STEERING AND POWER

We want a robot of reasonable size. So we choose a triangular base frame strongly constructed to hold 50 to 100 lbs. It can be made from aluminum angle iron or tubing so that it has strength and rigidity but is in itself light weight. We choose the triangle shape because we will power only one wheel, which will also be turned for steering. Fig. 10-1 shows the basic structure.

The actual dimensions of the base might be from one foot per side to two feet per side. Probably the wider it is at the base the less likely it will be to turn over, but then again you don't want it so wide as to be bulky and an obstruction in the house, where it will probably be demonstrated. You can probably find a size suitable for your plans somewhere in this range.

Although we are considering a robot frame that can do something useful, you might want just to build a scaled down version. You could make it perform as completely as possible as a model robot, and once developed could always be duplicated in larger form. This would be cheaper, use readily available toy motors and gears, and flashlight batteries. But regardless of size the concepts will be the same. We will try to point out any exceptions as they arise.

286

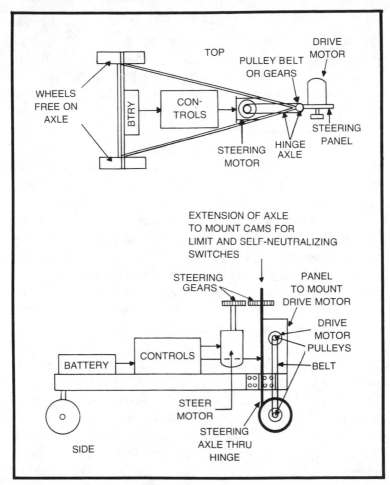

Fig. 10-1. A basic structure for a robot.

In Fig. 10-1 you see one idea of how you can use a panel on which the drive motor is mounted, secured to the frame with a strong hinge so the panel can turn left and right. This also shows how the steering motor might be mounted to the frame and geared to the steering panel so it can turn the panel left and right while the drive motor is operating. In the illustration the drive motor is connected to the front wheel by means of a pulley and a belt. You might use a gearing arrangement instead. The main idea is that the drive motor turns the front wheel, and the steering motor turns the whole drive assembly to cause steering.

Now with two leads from each motor connected as shown in Fig. 10-2, you can arrange a pair of switches to test out the unit. Notice that these switches are of the type which make no contact in the neutral midposition. When pushed to the left (up) they close two sets of contacts and when pushed to the right (down) they close another pair of contacts. The switches are double-pole, double-throw normally open types, and the two are identical. They need only be large enough in contact amperage to handle whatever current your motors will draw from the battery supply. A common battery supply may be used for both motors if it has sufficient amp-hour capacity.

Now you may play with the machine and see how it goes. If you close the drive switch for forward motion the platform should move forward at whatever speed you have arranged by the pulley gear reduction. You can check this to see, if it seems to move too fast, too slow or just right. You might reduce battery voltage if it is too fast, or increase it if it is too slow. You must have sufficient gearing reduction to carry the total weight to finally be placed on the frame.

Get some rocks, logs, or bricks to make up a weight and see if this can be pulled around smoothly. You might even use a small child, under 50 lbs usually, but larger if you want a larger robot. They will enjoy the ride while you are testing. The lady of the house will be gloriously happy that you have the children so occupied.

Start it. Stop it. Make it back up by putting the switch in reverse. Try steering it around and see if the steering is sharp enough. See if it has good balance in running and steering while carrying its load. If not, then increase the size of the base frame triangle. You'll have a lot of fun just making this much of the robot work the way you want it to, smoothly and quickly, with adequate turning sharpness while maintaining its balance and rigidity. Ultimately you can replace those human-controlled switches with relays or solid-state circuits. But even as described thus far you will find a challenge to your mechanical ability and your electrical know-how to make it all work.

Another word about the control switches. These are spring loaded to return to center, or neutral, off position. These are the best to use, for steering at least, as you will find that you will steer the assembly by closing the steering switch for only a second or so at

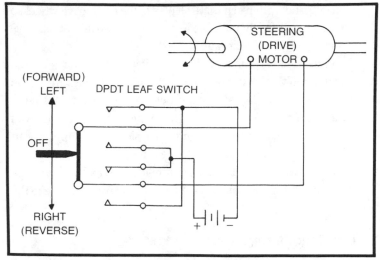

Fig. 10-2. Wiring the drive and steering motors. This circuit works for either.

a time, unless you are making a complete turn. The rest of the time you'll be trying to make the robot base go straight. When the steering wheel is electrically turned it will stay in the new position when the switch is returned to neutral. You have to send it an opposite turn command to get the wheel straight again.

THE DRIVE MOTOR

One electric motor operating on 12 volts which might be usable as a drive motor is the model airplane engine starting motor sold at hobby shops. They run pretty fast, but you can gear them down. Other sources of motors are 12-volt fan motors for automobiles. These can be found at junk yards.

EVALUATING A STEERING PROBLEM

In all probability you will have some problems trying to steer the base frame. You will find it difficult to find that neutral position of the front wheel for straight go ahead direction. Thus you will be required to make corrections constantly. The faster it goes and the faster you try to steer, the more difficult it is to accomplish quick and precise steering.

Let us see what can be done to solve the problem. First you must try to eliminate all play in the gearing of the steering motor to the steering panel. Then eliminate all play in the front drive wheel so it turns easily and freely on its axle but cannot wobble at all. This could cause a change in direction.

Next we need to think about providing limit switches on the steering assembly so that when we close the switch for a turn, the mechanism will move but stop itself before it reaches the physical limit of the steering assembly. If we have this happen electrically in both directions it makes our steering job easier. We simply have to push the steering switch one way or the other and hold it there. We do not have to worry about releasing it before the motor drives the steering panel into a mechanical bind as probably was the case in the beginning.

The Limit Switch Arrangement

We can mount two spring contacts as shown in Fig. 10-3 on a plate which has the steering axle through it. There will be an insulated cam shaped about as shown. This is fastened to the steering axle so that when the steering motor rotates the steering shaft to the right or left, this small insulated cam attached to its shaft will cause these switches to open, and thus remove power from the steering motor circuit preventing it from going further in that direction. Since the other limit switch would be closed, power can be applied to the steering motor to cause it to reverse direction. Thus we can have it go back to where it was or even in the opposite direction quite easily by simply pushing the steering control switch in the opposite direction. Study the diagram and follow the current flow through the switch contacts when the two upper ones are closed and then when the two lower ones are closed to the armatures.

The Straight Line Steering Problem

So we have solved one problem to prevent damage to our mechanism by having the steering motor jam the steering panel to the left or right in case we hold the steering switch closed too long at a time. We don't have to worry about that anymore. But we do have a second problem. This is our ability to steer in a straight line. As it is we have to hunt for the straight-line, neutral-position of the drive wheel. And sometimes that may take some doing. We get off a little to one side or the other and try as we may we cannot hit that

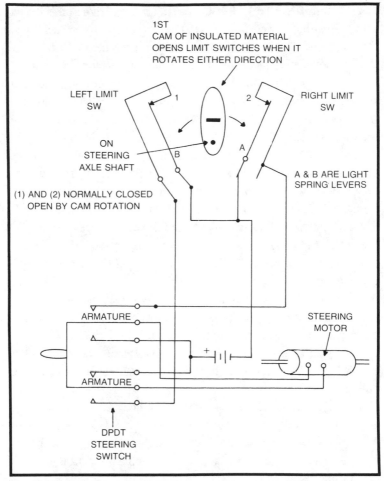

Fig. 10-3. Limit switches for a steering assembly.

straight-line position exactly unless we slow down the steering motor so much that it takes a week to move the steering wheel at all.

So what we want is an automatic neutral circuit which will make the front steering panel come to the straight-line position automatically in the absence of any steering command from use. Our steering will be easier then, at least much easier than it is now. And later when we try to control the steering by path sensors, this automatic neutral will be an essential factor in our robot's success. So to get this feature we need a self-neutralizing circuit added to what we have now in Fig. 10-3. So let us examine Fig. 10-4.

The Self-Neutralizing Circuit. The objective of this circuit is to provide battery connections to the steering motor in such a way that when we have released the steering switch to neutral, both relays shown will open—or become de-energized. The contacts shown will be closed by the positioning of the steering panel neutralizing switch cam. That will make the motor run in such a direction that it will seek neutral which is the straight ahead position of the steering wheel.

Now, instead of having the insulated cam open the switch contacts as it does in the limit switch arrangement (which we still need), we now need another cam which will close a pair of contacts and hold them closed whenever there is steering to the left or right. The only position of these contacts and the cam where the contacts are not closed is when the steering wheel is positioned for exactly straight ahead movement. Physical positioning of the neutralizing contacts is necessary to accomplish this.

Thus we see that the circuits will so energize the motor that it will constantly seek to drive the steering assembly to neutral (straight) when the control switches are off. Note that now you will connect your control switch so that it operates one relay in the left position and the other relay in the right position. We discuss the operation of Fig. 10-4.

Circuit Operation. It is simplest to show the operation of this circuit using the leaf contacts as indicated at points C and D of Fig. 10-4. These switches are open only when the steering assembly is at neutral, and one or the other is closed the moment the steering shaft turns right or left.

Look at the side represented by relay 2 (right), the batteries, and neutralizing switch D. If we cause relay 2 to close by applying power to its coil through the steering switch (this circuit portion is not shown), the negative terminal of battery A is applied to one motor lead via the relay armature and the other motor lead is connected directly to the positive side of this battery by a direct connection at X. The motor will now run. And *if* you have arranged the motor leads properly the steering shaft will turn, causing the cam to immediately close switch D. This doesn't have any effect because the switch is isolated from the motor and batteries as the upper contact of relay 2 is *not* touching the relay armature when the relay is energized. So there is no effect from the neutralizing switch, and the steering motor will continue to turn the steering shaft until limit switch E breaks the circuit.

Now consider that we put the steering switch in neutral so that relay 2 is de-energized releasing its armature to close the upper contact, which is connected to the neutralizing switch. If you trace the power flow now, you will find that the neutralizing switch has connected the motor lead to the positive terminal of battery B. And since the other motor lead is connected to the battery negative at X, the motor now runs in the reverse direction. It will keep running until the cam is centered, and neutralizing switch D opens to break the

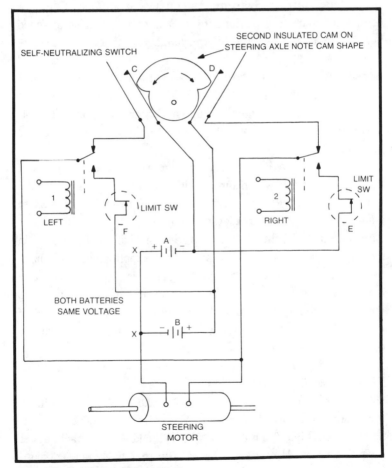

Fig. 10-4. A self-neutralizing circuit.

circuit. We have automatically caused the steering motor to seek neutral. Check the opposite side of the circuit and you'll find the same kind of action takes place.

The only critical area of concern to this circuit is the neutralizing switches. Make sure they are so connected to the batteries that rotation in one direction will close the proper switch to cause the motor to reverse direction when the relays are de-energized. If the motor runs the arrangement in the wrong direction, reverse the motor leads. Notice where the limit switches are wired in this diagram.

A One-Battery System. As you investigate this system for controlling your robot platform by means of a wire cable attached between it and the switches in your hand, you will begin to think of how to proceed to make the robot a self-guided, programmed machine, which is what robots really are—aren't they? As you study Fig. 10-4 you will begin to wonder if, in fact, you might not use just one battery—large enough to handle all motor drives and relays. It would require a reversing relay arrangement instead of the two batteries shown.

Yes you can do this. We have shown two batteries first because it is simpler to follow a diagram with two batteries. Actually the two-battery arrangement works quite well. It might be adaptable to a smaller robot using tiny electric motors and relays of the 3-volt variety. Small nickel-cadmium batteries similiar to those used in model airplane radio-control systems can supply power. We have used relays to route power because, again, it is simpler to analyze a circuit with them. And they are available in various voltages and contact arrangements from such supply houses as Radio Shack. If you are electronically minded you can redesign these circuits to operate from solid-state components.

It is quite easy to reverse a permanent-magnet electric motor by using a small relay with a set of contacts such that there are two poles and four contact connections. A double-pole, double-throw relay is their technical description. We need one to use as a reversing relay. On Fig. 10-5 note poles 1 and 2 and their contacts of the reversing relay. Also see the use of the two single-pole, double-throw relays (1 and 2) which we used in Fig. 10-4. We also still use the limit switches and the automatic neutral switch arrangement of the previous diagram. Study Fig. 10-5.

Let us trace the operation. Assume that we close steering switch Y to give a left command signal. This causes the left relay (2) to close its pole to contact (6). And since the limit switch in series with this line is normally closed, we send power from the positive

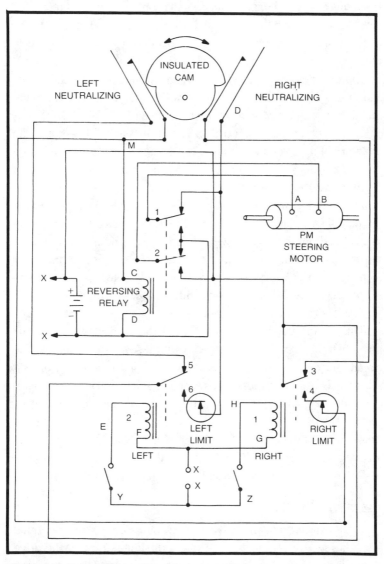

Fig. 10-5. Self-neutralizing with a single battery.

side of the battery through contact 6 to the top contact of the reversing relay. The line to the right neutralizing switch is also energized but since this switch is normally open and will remain so for a left movement of the cam, nothing happens on this line.

At the reversing relay, however, power now flows through pole 1 to motor lead A. Motor lead B is connected to pole 2 of the reversing relay, and this, through its top contact, is connected to the reversing relay, and this, through its top contact, is connected to the negative side of the battery. So now the motor runs. And if we have the motor leads to the correct poles (reverse them if they are not), the motor drives the cam left and will finally open the left limit switch and close the left neutralize switch. So this is how everything remains as long as we have switch Y closed.

Now let us place control switch Y in the off position. The armature of the left control relay (2) will engage contact 5, and since the neutralizing switch is now closed by the cam, power is routed to the reversing relay, so it becomes energized. This reverses the polarity of the battery voltage applied to the motor leads, and so the motor will now run in the opposite direction to drive the cam to the neutral positon (the steering wheel also is now in neutral). At neutral the left neutralizing switch opens to break the circuit and the motor stops. Now the steering wheel is positioned for straight ahead movement, and all is ready for another steering command.

Notice that now we give signals by simple closing switches X or Y. This means that we can use any of the path control systems previously described to operate relays at these positions to close the circuits for path steering of the robot. And, in the absence of path steering signals, the robot moves straight, and that is what we want.

Notice again the location of the limit switches in this circuit. Recall that these must be physically located so that the cam will cause them to open at each end of the limit of arc you set for the steering wheel rotation. You can set this arc according to your desires for how sharp a turn you might need or want.

THE DRIVE SYSTEM

You can operate the drive system by using a voice operated relay or some other means. We might think about using the drive motor to operate a crank and linkage, such as we saw and learned about in an earlier chapter, to made a robot walk. Your drive motor might operate feet this way if you want to investigate a walking robot. But you might also consider that making a turn with a walking robot is a very difficult thing to accomplish. You might have a go at it anyway.

Any small permanent-magnet motor strong enough, through its gear train, pulley system, or chain-drive system can move the steering arm. And if it can do it with a load, you have one big enough. It has been suggested that a visit to the automobile parts shops or junk yards will produce small motors used in cars to raise and lower windows or move seats or operate air conditioners. These have built-in gear trains which perhaps will fit your needs. They are reasonably fast and quite powerful.

HANDLING THE NEUTRAL-SEEKING OSCILLATION

We need to mention something about the speed of response of the steering motor and its hunting when seeking neutral. If we have a motor which drives too fast, there might be some overshoot which could cause a slight amount of hunting if the neutralizing switches are placed too close to the cam. Of course the overshoot will immediately be compensated for because each neutralizing switch will work to cause the arm to center. You want these switches as close as possible to the neutral position to eliminate a dead zone. But, really, you don't want any overshoot if you can help it. You might have to sacrifice a little steering speed to get smooth action without hunting. Or you might tolerate a little hunting in order to get a fast response. You might consider that the robot's "clothes" will cover the jitter movements of the front wheel. The "clothes" will probably be a panel of aluminum or a cone of aluminum or light gauge steel. This will protect the insides and also keep peering eyes from seeing what makes it go. But remember that hunting can cause the battery to run down sooner, which you probably won't like. A slower response and no hunting is best.

MORE ABOUT THE DRIVE MOTOR

At this stage of development you have a robot platform which can be controlled by cable and which can be steered and moved backward and forward by your commands. You do not have a vari-

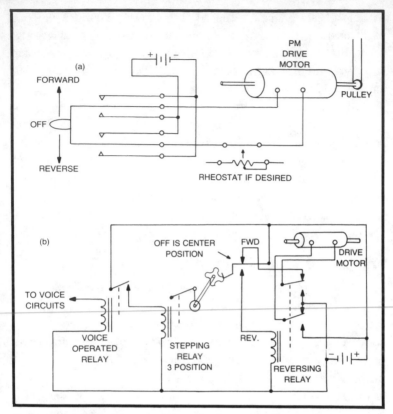

Fig. 10-6. Manual (a) and voice-operated (b) circuits.

able speed of motion. Figure 10-6(a) shows a rheostat which you can vary to give some speed control if desired. We do not recommend using taps into the battery to get less voltage to the motor. This will run down some cells faster than others, and perhaps ruin your battery. Actually we suggest that you use gearing or a pulley arrangement which will give one speed. Use that for a while as you experiment on various other parts of the control system. Then, later, you can examine speed control concepts and make some changes in your robot's speed if you so desire.

DRIVE MOTOR CIRCUITS

We have not yet presented a diagram of the wiring for the drive motor and do so now in Fig. 10-6 (a) and (b). As you see at (a) a double-pole double-throw switch can be used. It reverses the battery voltage to the motor when it is moved from forward to reverse. In the center position there is no contact made to the movable leaves of the switch, and so the motor doesn't run. It is a quite simple circuit and a good testing circuit. At (b) we show how you could operate the drive motor from a voice-operated relay. One sound might be forward, the next sound reverse, and the third stop. This sequence would repeat over and over. Finally, there is a simple arrangement using a reversing relay, as shown in Fig. 10-7.

Probably you are wondering if the drive motor could be controlled by a light operated relay. The answer is yes. It might also be controlled by radio commands using a radio control system of the types currently available. Steering could also be controlled by radio.

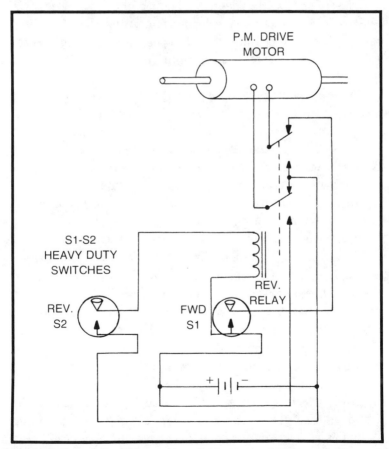

Fig. 10-7. Using two switches and a relay to reverse a motor.

But before we get to involved in this at this time, let's look at a simple programming concept which might be fun to instrument.

SIMPLE PROGRAMMING FOR THE BASIC ROBOT PLATFORM

It is fun and interesting to see what actions and motions you can make your platform accomplish. This is basic to having your robot do other tasks which you might assign. Let us consider the first proposition, which might be to have it follow some kind of a designated, but not prefashioned path. A prefashioned path would be a buried wire carrying an AC signal, as one example. We will consider the designated path to be one which is defined only by time and distance, and turns.

Now, if you will go back and refer to the program drum or motor driven series of switches described in Chapter 5, you will no doubt see how to apply such a unit to do the switching which we have been doing by hand. After all, are we not just opening and closing switches to start and stop and steer the platform? Yes, we certainly are doing just that.

If we consider a program of opening and closing contacts and/or switches then we can imagine sending our robot platform on missions to go to certain locations and return, according to how we program the drive and steering switch closing and opening apparatus we have been discussing. Of course we are now considering that the switches will be closed and opened by some device such as an electric motor, and that we can adjust it so that it opens and closes the various switches for various time lengths. If we do this we will be causing motion of the platform as a function of time.

Let us imagine that we will start the robot on a straight course, say along one side of a room. We know how long it will take to get to the end of the room because we can time this with a clock. At the end of the room we will want the steering section to be energized so the robot will make a 90 degree turn. Then the platform is to continue along the end of the room for a specific time, make another 90 degree turn, etc. until it gets back to where it started.

Let us imagine that you, with the help of someone to read a clock, will actually steer the platform through these maneuvers the first time to get the times required for each segment of the trip. When this is done you will have a chart which will probably read something like this:

1. Straight for one minute 10 seconds.
2. 90 degree turn. Steering for 10 seconds left.
3. Straight for two minutes and 5 seconds.
4. 90 degree turn. Steering for 10 seconds to left.
5. Straight for one minute and 10 seconds.
6. 90 degree turn. Steering for 10 seconds to left.
7. Straight for two minutes and 5 seconds.
8. 90 degree turn. Steering to left.
9. Stop.

Now all you have to do is to arrange the contacts on the drum or the switch actions of the cams. Then start your robot platform out and see if it actually can go around the room as you first steered it, but this time on its own.

You will find that you will have to make corrections and adjustments. So we recommend that you always, during this experimental stage at least, have a master switch which you can throw to the off position to quickly disconnect all power. This just in case the platform does not do exactly what you imagined it should do. Then you make the necessary adjustments and try again. When finally you have a program that will work for this room, it would be nice to be able to remove it (the drum for example) and put on a new drum which could steer the robot around the next room in the house. Presently you will have programs for each room. And even some for outside trips, such as down the sidewalk, turn around and come back. Of course this presupposes you always start at the same place, but that should be no problem.

If you plan to have your robot go out into the yard, remember that its wheels must be large enough so they won't become bogged down in grass or dirt. Think of this ahead of time while thinking of your platform size. Also the motor power must be large enough to propel the platform around the yard. All this can be tested using cable controls until you get that part of the system working properly. It may mean a change in the power drive. But that is part of the fun, the experimentation.

This programming is probably the simplest that can be used in a robot, and open enough so you can examine the works and see how everything fits together to make it go. It is possible to have quite elaborate programming using this system, even to doing real tasks. If the robot (after construction is complete) is given a sweeper, for example, it might even do a pretty decent job of sweeping the floor in each room—if your program is that detailed.

Now comes one question which you might have thought about, and which might be bugging you. That is, how do you get a gradual (large) turn when the steering section simply goes from neutral to left and back, or neutral to right and back. Under our present system there is no in between in steering so all turns are sharp and in the tightest possible circles.

Actually, in a final model we would like to have a servo with potentiometer feedback so we could program any size turn we wanted, but for now in this simple beginning we must overcome this apparent deficiency by controlling the time that the steering section is energized. You might have already checked this out using your cable controls. If you want to make, say, a 10 degree left turn, you simply give a very short left signal. The platform turns a little and then the steering wheel will neutralize. If you didn't get enough turn, you give another quick signal to left. If you gave too much of a signal, you have to reverse direction and give a very short right command.

Well, in the programming, the platform is not going to have any way to know if it turns too little or too sharply. This is a bad feature of this system. But we can approximate the gradual turns by doing just what we said, causing the steering switches to be closed for only a brief period of time as the platform is in motion. Check your timing on this with a stop watch and you might be able to program a pretty good gradual turn if the platform doesn't move too fast and the steering motor doesn't cause the steering to be too fast. Try it. You'll have fun with it. And there are so many things the robot might do using just the 90 degree turns that you will have much experimentation before you exhaust this kind of possibility.

It is now possible to consider completion of the robot by giving it a body, arms, and head; or you might just want to advance to a more sophisticated type of motion and steering before going ahead. Since we have had much discussion of methods and ideas of more sophisticated systems in earlier chapters we will go ahead and examine what might be considered to be the next requirement of a basic robot, some simple arms and their movements. This will take some doing to get them moving as they should, both from a mechanical and an electrical standpoint. So, now, if you are pleased with what you might have built and tested so far, let's march forward.

ARM CONSTRUCTION AND OPERATION

Let me tell you this is an area which scientists and engineers have been exploring of late with great diligence. As you know from previous information in this book, there have been in-depth mathematical studies to develop algorithms (mathematical formulas) which will enable those in the know to build arms that operate much like our human appendages. This is a complex subject because we are not just concerned with a simple movement of one part, but we have to consider the bending and twisting and turning and shortening of various parts of the arm if we are to make it act realistically. Then too, there is the really big problem of having two arms which can work independently, yet coordinate together as our arms and hands do.

Wanna look at one arm which can do lots of things in a truly marvelous fashion? Then look at Fig. 10-8. Notice the various parts which are movable by electric motors. There is the pincers hand which is so arranged that it can grasp large objects with a direct closing of both sides. Note that it does not tend to be just a clamp pivoted at the center open and closing like a V might if hinged at the bottom.

Then there is a wrist motor which gives rotation in either direction to another motor section which permits the hand to move up and down in the view shown. Examine your own wrist and hand action and see how closely this arrangement is to your own hand and wrist action. There is a hinge at the elbow section, farther back, and another at the shoulder position, which is mounted against the submersible chamber itself. As you can imagine, when the shoulder motor causes the arm to move in an upward direction, if, at the same time, the elbow joint causes a downward movement you will get an extension of the arm to increase its reach.

Again, work your own arm and elbow joints and check the action against the arms capability as shown in this figure. General Electric, who developed the whole thing says it is in use for underwater exploration and devices manipulation. It can even fasten things together under water. It is operated by a human operator who has an attachment which fits his hand and arm such, that as he moves, so moves this robot arm extention. In a true robot we want such arm movement controlled by the robot itself.

Now there might be a combination of power sources for such an arm. You might use hydraulic pistons linked to cams and levers to cause up and down or rotational movement. Pneumatic (compressed air) might also be used if you take the proper precautions, but electric motors with proper gearing are generally used at each joint in the system, and they have feedback potentiometers attached to the output shafts so that accurate servopositioning can be accomplished.

Your wires might be run back through hollow tubes of aluminum or thin steel tubing or along the inside of little I beams inside the arm sections. Of course the arm shown in Fig. 10-8 is designed to withstand extreme pressures of water, as it is to operate at ocean depths. We don't have to be concerned with seals as they must. But, referring back to Chapter 1, you might see how such an arm can be covered with plastic to give flexibility and a nice appearance, yet be removable so that all parts requiring maintenance can be exposed.

That is important in building and designing a robot. Don't box yourself in such that you have to disassemble the whole robot to get to some section or part which might fail or give trouble of some kind. Plan your robot's construction so that you can gain access to each and every part and cable with relative ease. I like that word relative. That means with only as much trouble as you are willing to expend to get to that part.

But, back to the arm. Examine Fig. 10-9. The drive mechanisms of the system are operated with electronic controls and employ permanent-magnet servo-motors with attached planetary gearheads. One motor gearhead combination is used for the shoulder and elbow motion, the other is used for the wrist motion. The potentiometers are packaged internally for protection and to minimize the profile. They are spur-gear driven off the motor shaft.

Two main sets of counterweights are included with the system. They extend behind the shoulder approximately 8 inches (20 cm). The sum of the two sets eliminates the shoulder joint imbalance.

Fig. 10-8. An underwater robot arm. Courtesy General Electric.

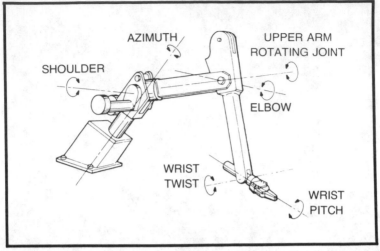

Fig. 10-9. An advanced action manipulator system arm. Courtesy General Electric.

Adjustments of these counter-weights between the right hand and left hand manipulators provide balance about the tilted azimuth axis.

The servo controls include an operational amplifier output stage to drive the motors. A power conditioning card feeds positive and negative 8.1 volts to the potentiometers and positive and negative 6.2 volts to the operational amplifier. In addition, a lag network (remember our discussion of rate circuits and why they are used?) provides servo compensation (and eliminates hunting). Finally, gain K between the master and slave is controlled by a stepping rotary switch which changes the master (control) input resistance to the operational amplifier.

THE POSITIONING PROBLEM

We mentioned a servo amplifier, a comparison unit which can compare the input command of one polarity to an output (feedback) signal of the opposite polarity. When the second cancels the first—due to proper positioning of the output member—then the output of the amplifier does not exist and so there is no drive power whatever to the motor control devices which may follow this unit. Comparison amplifiers in solid-state units are availiable at radio and parts stores and could be used in these applications, and should be used if possible. They can be followed by whatever kind of motor driving

306

system—relays, power transistors, or whatever—that you may want to design into your own system. One such operational amplifier is the LM 308, or equivalent.

When positioning by remote control (such as cable control of your robot) any part of an arm such as was shown in Fig. 10-9, you must have some observation of the position of the arm and wrist and hand. Thus you become a part of the feedback. It gives that position to your brain which controls your hand. This, in turn, controls the mechanical arm. So you juggle for the proper positioning, or you may have an electrical feedback which will cause the arm to move to a precise position specified by the input voltage.

If you have an input voltage and a feedback voltage then you can probably send in a command and watch the arm position itself according to the voltage signal you have initiated. Realize, however, that a subprogram could also cause the arm to respond to its signals, or an input sensor might generate an input voltage which will cause the arm to do something. When we say arm we mean whatever part of the appendage is moved for the purpose required. It may be the arm, the forearm, wrist and hand, or fingers. If it is fingers it might be all of them, or just one (if you have so designed your robot).

Many times when trying to imagine the building of a simple experimental robot in the home workshop, or sometimes in a laboratory, one is confronted with the situation wherein we know what we need in mechanics or electronics, but don't know exactly how to get it, or how to do it, or how to instrument it to make it work. Let us do a little experiment here with the simplest kind of feedback system we can imagine to give us some variable positioning of an elementary robot arm so that we can learn more of the problem of moving it as we might want to move it.

We choose one joint to move, the shoulder, knowing that we might instrument the other joints in this same way. And as we do one joint we will learn what to do for the others in addition to giving our robot platform at least one arm movement, even if, at first, we have to keep the arm stiff and have it move only up and down. Let us examine Fig. 10-10. You may have to experiment with this basic circuit to make it work as we will outline. Here we are going to use a simple transistor operated relay system and control the voltage input to two transistors from a bridge circuit of two potentiometers.

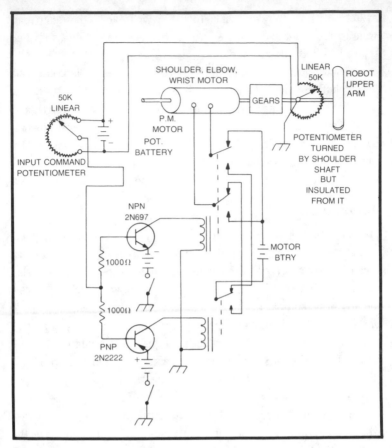

Fig. 10-10. A relay servo system with feedback.

This circuit is the type we discussed and examined way back there in a previous chapter. The relay circuit is basically the same as we examined in our platform steering section except that we have now made it more sensitive by adding the transistor premaplifiers.

As shown, one transistor is to respond to a negative voltage input and the other to a positive voltage input. This is required because the potentiometer bridge will give that kind of output. Of course when the bridge is balanced, there will be no voltage difference between the potentiometer wipers, no input to the transistors, and the relays will not be energized. The relays must be sensitive enough to be operated by the transistors, and most radio parts houses have little circuits showing how to operate relays from transistors, giving the relay and transistor types and other technical information.

One relay is our reversing relay; the other acts merely as an open or closed switch so that if the polarity of the input is, say, negative, then the switch closes and the motor starts running, moving the arm and feedback potentiometer wiper. Of course both are moved through a suitable gear arrangement on the motor shaft. Now the feedback potentiometer arm must be moved in such a direction that it balances the bridge. If the arm is moved in the wrong direction it will continue to move until it reaches its mechanical limit and then will still try to move, causing the motor to get hot and cause all kinds of trouble. Reverse your motor leads if this happens.

If the wiper is moving in the right direction a small movement of the input potentiometer will cause a small arm movement, and so on. You will have built and made operational the simplest kind of servo system but one which will do the job for you, at least in the beginning.

Now, to consider the next advancement we begin to think about a system which can use a voltage input to make the arms move

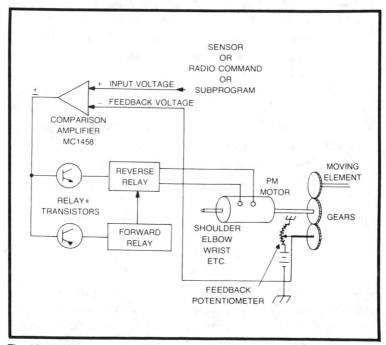

Fig. 10-11. Use of a solid-state amplifier in a servo comparison.

instead of the physical rotation of the input potentiometer we have shown. Let us now look at Fig. 10-11.

The comparison amplifier takes as its input two voltages of opposite polarity and gives an output according to the difference between them and of a polarity which is the same as the largest of the input voltages. It thus becomes ideal for what we desire here. We can supply the input voltage from any source—subprogram, radio system, or sensor—and the feedback will be the second voltage to the amplifier. The motor will operate in accord with whatever voltage is the largest and in a direction dictated by the polarity of the larger input voltage. We can use relay switching as was shown previously to control our motor, or we might want to make our motor work without any relays, instead using transistors. We'll give you one such circuit in the next chapter, but we call your attention to TAB book 841 "Build Your Own Working Robot," and to TAB book 135, "Radio Control Manual." Both of these books contain more of the circuits in them.

But here, with relays, you have an arm positioning system. It probably won't be easy getting the gears and the motor all arranged properly (unless you use one of the car-window or car-seat motors we talked about earlier which have the gears built in). You also have to get all the mechanics and electronics together such that it all works. You will probably work in this second project of your robot construction after getting the steering and motive system all working properly. But then you might program the arm, using the drum with contacts, or the drum with risers and switches in a time arrangement. You might be able to send your robot to the front door, have it stop there, wave its arm (arms) and come back to you. By the way, just in case you might need a good circuit which is sensitive to small input voltages and will operate a relay, we include one in Fig. 10-12.

You can spend a lot of time on the arm system, expanding it as you see fit, making two arms which can be energized separately and independently (as our human arms do). And then if you want a real challenge, try devising a system wherein the robot might use both arms independently to, say, grasp something with one hand and then perform an action on that something with the other hand. Like the Mobots shown earlier, pouring from one flask into another.

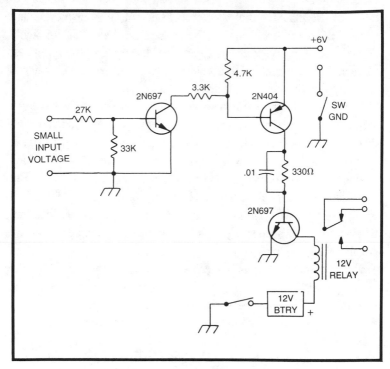

Fig. 10-12. A sensitive relay amplifier circuit.

We might spend a moment thinking about the hand. It has been found that a good hand can be made of a ½-inch layer of plastic inside the metal clamp. This can then deform to encompass different objects and surfaces, like our flesh bends and stretches to accommodate various objects. Also, don't forget that the pincers type of hand is really not the best kind. Your mechanical hand should be slightly curved, if you do not have movable fingers, and have sides which move parallel to each other in closing and opening like the jaws of a monkey wrench do. If the inside is then plastic or rubber coated so it can have a little give, you will have hands which can grasp most objects and handle them reasonably well.

HEAD MOTION AND "LIFE" INDICATORS

Once you have progressed beyond the arms, elbows, wrists, and hands then whether or not you want a head which moves is your choice. It is not necessary and lots of robots just don't move their heads at all. Plastic bubbles, parts of balls of large diameter, and other shapes can be used as a head. You can have flashing lights inside, or flashing eyes, or whatever, which are programmed and/or voice operated, either by your voice or someone else's. This can add to the fun of having a robot around the house.

If you want to program a servo system controlling the head you might make it respond to a sound such that it, with its sound sensors (like our ears) might make it turn its head toward the source of the sound. Of course, normal voice tones disperse so fast, and over such a wide area you probably won't have much success with this. But don't give up trying. You might find a way, just as you might improve on all the ideas we have presented throughout this book. After all, that is one of our prime purposes: to stimulate *your* imagination and genius. Now you have some basics. The improvements are next in order, and improvements are called advancements. This may take lots of study and investigation to find out the hows, whys, and wherefores of making a better robot. But, as they say, build a better robot and the world might beat a path to your door. At least the robots might do it.

I have just consulted with my close friends and associates, the gnomes and elves and trolls and magicians and wizards who dwell in and about my desk. And they have arrived at a (miracle for them) consensus as to what we should do next. I warn you, their recommendations are not always easy to follow, but we will do the best we can. How about joining us in this next chapter?

Chapter 11

Radio Control of Your Robot

As we indicated earlier it is possible to control your robot by radio signals sent via a transmitter to a small receiver inside the robot. The receiver sends the signals to various servomechanisms which can then control arms and head movements and operate switching devices to give other actions. Of course you can steer the robot this way, or you might transfer control over to a subsystem which will take the robot on a planned route if you so desire.

A TYPICAL RADIO CONTROL SYSTEM

Let us look at the components of one radio control system currently available and ready for installation (Fig. 11-1). There are other types available, even as kits from Heathkit. This system was designed for use in model airplanes, cars, or boats, but will adapt easily to robot control; the robot is just another form of these kinds of machines, more elaborate, true, but still a generic descendent.

One nice part of this radio control gear is the servo-mechanisms, which are small or medium in size, powerful and fast, having operational amplifiers and feedback which gives very precise positioning capability. You can obtain up to eight channels of control. Since only one channel is required for a drive motor and one channel for steering, you have six channels left for functions such as arm, elbow, and wrist movements.

313

Actually you might want two radio control systems to instrument all the possible appendage movements. But you may be able to combine some motions if you develop mechanical linkages which, for example, will extend the arm as the shoulder rotates it forward, and retracts the arm when the shoulder rotation is back down again.

Of course, if your arm is simply a rotation up and down and the hand on the end is a simple open-and-close arrangement, then you don't need a wrist or forearm movement. And so you can reduce the number of channels necessary. All this will depend upon what you want your robot to do. That will take some planning. Make a list of everything that you would like to have your robot do *someday*. Then work toward that accomplishment. We have another idea for the use of the major part of a radio control system in a robot. So after we find out a little more about the system, let's see what we can do with it in this automatic sense.

How it Works

A radio control system such as we have shown operates with a series of pulses which carry information for each individual channel. Thus for an eight-channel system there are eight information carrying pulses, and the position of each of these with respect to a given starting time for each series determines what kind of information is being transmitted and received. There are two more pulses required in each train, a starting pulse and an end pulse, or there must be a time delay between each train of pulses transmitted to enable the receiver decoder to get set again to receive the information pulses in the proper sequence. The pulses are transmitted serially. When they are received a timing signal insures that they will be routed, one to each servomechanism, in much the same manner that computer pulses are routed to their various destinations.

It is possible to design a system with 16 channels of information. Two eight-channel systems would be more expensive and slightly larger than one system, but this may be necessary in some cases. If a need someday arises for a 16-channel systems, the technology is available and manufacturers will undoubtedly expand their systems to supply as many channels as people want. It is important to realize that all channels, through serially operated, operate at such fast rates that for all intents and purposes you can consider that you have

Fig. 11-1. A radio-control system.

instantaneous and simultaneous responses from all channels. They operate without causing any interference, one with another. The eight-channel systems are proven and readily available.

WHERE THEY OPERATE

As with the CB channels, there are specific channels set aside by the FCC for radio control. No license exam is required for channels in the ranges 26.995 to 27.255 MHz, and 72.08 to 72.96 MHz. An amateur radio license is required to operate in the range 53.10 to 53.50 MHz.

There are thousands of hobbyists who fly model airplanes, race cars, and sail model yachts using radio control systems. There is usually enough separation between sites where the models are operated and such excellent control of transmitting frequencies at these sites that no interference is generated on the normal outings. On the 27 MHz band it is possible to get some CB interference, and this could be a problem. Most commercial radio-controlled toys operate at 27.255 MHz as of this writing.

Since a robot will usually be close to the operator it is not unlikely that a much higher frequency channel could be used (if the

FCC approves it) for robot control. This might be at, say, 465 MHz. At this frequency a highly directional antenna would be used, which would reduce interference, and make the RC elements of the robot system so small that they would be almost inconspicious. A small loop on the receiver would be all that would be necessary to receive the signal. A small directional beam antenna on the transmitter would occupy only a few inches in width and length above its box.

CAPABILITIES AND LIMITATIONS

The capabilities of the hobby radio control systems now in existance are many. They work. They use small nickel-cadmium batteries. They have good range. They are rugged, small, and reliable. They are easy to install and require very little maintenance except to keep the batteries charged and properly cycled when necessary. You must cycle nickel-cadmium batteries or they will develop a memory as we stated back in the chapter on primary power for robots. If they get this memory they won't work as long as they should, nor as efficiently.

The limitations of radio control systems are that they are somewhat expensive and they may be subject to interference from various sources. Batteries must be kept charged. And if the system has relays built into it—most do not—they must be kept clean and the contacts bright. The relays must hold their adjustment, which sometimes gets joggled during rough motion from some accidental means. When flying a model airplane, interference might cause the loss of the aircraft. When handling a robot, probably the only indication that there is interference will be when it starts doing things you have not told it to do, such as wave its arms, yell, turn, stop, start, etc.

ADAPTING THE RADIO-CONTROL SYSTEM FOR ROBOT USE

Radio-control system servomechanisms are designed to work with pulse inputs. So they are readily adaptable to units which might work with a small computer-generated series of pulses instead of having the pulses sent by radio. Or you might want to have both sources of inputs, the radio and the computer. Since this modification is a definite possibility and the basic equipment is available, this becomes a most attractive possibility for our robot's control system.

The radio link consists of a transmitter modulated by a coder which provides pulses. These are transmitted over the equivalent of a single wire to the receiver. From the signal, the pulses are separated and then sent to the various servos to cause them to operate. Each servo has proportional control, meaning that it can move as little as desired or as much as desired. The servos come in a variety of sizes, shapes, and strengths, from subminiature to the large, standard size. Figure 11-2 shows a subminiature servo from Heathkit.

Fig. 11-2. A subminiature servo showing the motor, amplifier, and gears.

Suppose we take such a control system and simply bypass the radio transmission and reception part, and make a direct, coaxial wire connection between the pulse coder and the pulse decoder of the receiver. Of course the servos would remain connected to the receiver pulse decoder section. The input to the transmitter coder, instead of being the movement of levers and switches, might now be voltage(s) which are superimposed on the voltage generated at the wiper of each potentiometer fastened to the levers and switches. Notice Fig. 11-1 again; these levers and switch-like controls actually move wipers of potentiometers which in turn control the individual pulse timing.

The voltages superimposed on the potentiometers can come from sensors such as photocells, microphones, amplifiers for magnetic coils, or even programmed voltages from a central programmer which might have a small computer memory. In general, since the potentiometers control the value of a single polarity voltage, you would only have to adjust it, then add to or subtract from that particular voltage to have your robot respond. The addition and subtraction process might be done with a comparison amplifier which algebraically adds its two inputs. Of course you might also do this with a simple resistor network if you concentrate your thinking on just how it might be done.

So let us assume that you have obtained a radio-control system which gives you eight channels. You have connected it as we suggested and now it operates various parts of your robot. One channel might be connected so that it responds to external commands through the radio link, if this is desired. The others would all work from sensors or internal program subroutines. You might want to get a completely built system, or you might want to get a radio-control kit and build it all yourself under such excellent instructions as are given with the Heathkit units. See our TAB book 825 "Flying Model Airplanes and Helicopters by Radio."

Let us examine in Fig. 11-3 a method of how we might instrument a robot's arm using these model airplane servos.

We have one servo which can rotate the shoulder joint. A second servo, using a push-pull linkage can move the elbow joint. This linkage is like that used in model airplanes for rudder or aileron control. Another servo built into the forearm can rotate the wrist like

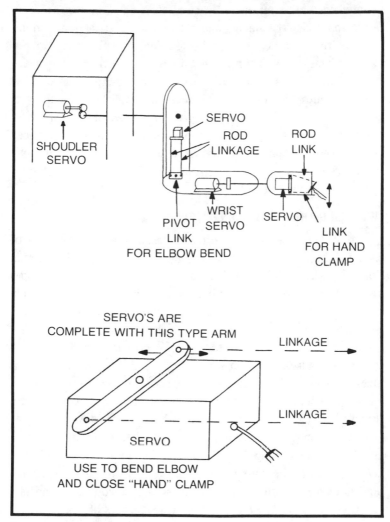

Fig. 11-3. Using servos to provide arm and hand motion.

the shoulder. And finally, another push-pull servo linkage can open and close a pincers hand section. Remember that the hand should be slightly curved and coated with a ½-inch plastic or rubberized material to enable it to get a good grip. You can probably devise a better hand closing method which will give you the parallel grip we discussed earlier.

Four servos are used for each arm, and that exhausts an eight-channel system, but fortunately we can add another eight channel system to get motion control. This can be done through a servo controlling the amount and polarity of electrical power to the drive motor, or by expanding the RC servo so that they drive a higher power transistor output section, which in turn might drive the motor. Another channel could be used for steering control. The steering control, incidentally, might be accomplished directly with one of the larger RC servos and then the other five or six channels of this second system could be used for robot functions such as voice, head movement, lights, and whatever. As a matter of fact you might incorporate even more eight-channel systems to realize the full potential of your robot.

ANALYZING THE SYSTEM OPERATION

If we consider the performance capabilities of this kind of system, then we can see what we have to do to get the internal controls working if we examine Fig. 11-4. Taking out the radio transmitter and receiver sections, which will not be necessary for internal control, we have some multivibrators, pulse amplifiers, and internal coding and decoding units, which are integrated circuits on most systems. The integrated circuits do the complex jobs of generating pulses and then separating the pulses into the proper channels so they are routed to the proper servomotor systems.

Each servomotor unit has its own small amplifier and motor driving circuit using transistors. No relays are employed.

An expanded block diagram showing what is in the integrated circuitry is shown in Fig. 11-5. Notice that the feedback potentiometer is geared to the output shaft of the motor and the comparison amplifiers are used to compare pulses so that all you need send into these systems is a command signal. This consists of a pulse which has a time relationship to a reference pulse which is also sent into the servoamplifier system.

If the command pulse leads the reference pulse, it will cause motor rotation in one direction. If the command pulse lags the reference pulse, then rotation of the motor is in the opposite direction. Matching of the pulses causes the motor to stop movement. The amount of lead or lag governs the amount of power applied to the motor.

When the motor shaft rotates it turns, through gears, the output shaft, which is also connected to a small internal potentiome-

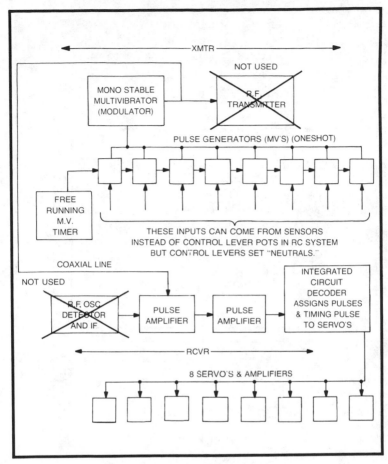

Fig. 11-4. Radio-control system with radio links removed.

ter (feedback pot.) which then adjusts the reference pulse to balance out the command pulse. When this is done the output shaft has moved the desired amount, and the motor stops running. If the command pulse is now caused to go back in time (lag), so that it has the time relationship for the output shaft neutral position, the reference pulse will not balance it out with the servo feedback potentiometer in its present position. Thus the motor will run, turning the potentiometer until balance does occur. Then the output shaft of the gear train will again be at neutral. Notice that all the feedback necessary for these operations is contained inside these small units which you can buy either in kit or completed construction form.

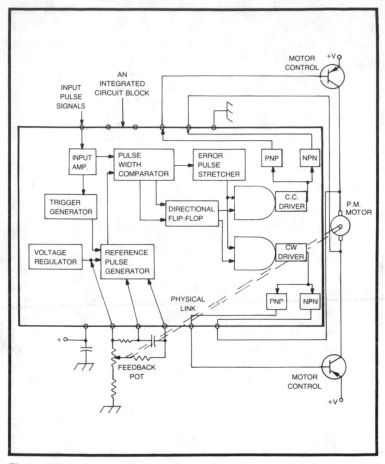

Fig. 11-5. Internal workings of a servoamplifier integrated circuit.

As you can see from Fig. 11-4, it would be easy to send in voltages to each of the command potentiometers such that these voltages would change the voltage at the wiper output. This will change the coder pulse timing and thus order a servo movement. If the voltage added is small, a small movement is commanded, and if the voltage is large, a large movement is commanded. This is proportional control.

So, thinking it over, we do have all the components already put together for us to make our robot very operational and highly intelligent. With two eight-channel systems he can respond to at least 16 command signals from sensors. That should be enough to do

322

about anything we desire to have him do, and if not, we can add still more systems.

THE COMPUTER INTERFACE

If our robot is going to be a learning type and if we want to really get involved with our project, then we now begin to think of putting a small computer inside it which can operate all these servos and subsystems which we have just discussed. Can we interface a computer with the radio-control system coder? Interface is that big word meaning can we connect them together. The answer is yes, it seems plausible.

The output of a computer consists of pulses. And we have bytes, which are units of 8 pulses, and words, which are many bytes. These pulses do not change position with time. Information is conveyed throughout the computer system by the existence or nonexistence of pulses in a fixed position in the pulse train, or byte. Pulses have a fixed polarity. They normally do not change polarity in a given system. So, now we can imagine how we might use this information.

If we have a means of connecting a group of gates to our radio coder, such that these gates can be operated by the computer clock, then we can supply the voltage to the coder from the computer in the proper channel at the proper time. The computer can get its pulse information from a memory bank, just as it does now, as a function of time and of its own subroutine program. Figure 11-6 indicates one interfacing section which might be developed. The gates might be such that each one opens the next, once the first one has been triggered by the clock. The timing, then, of the computer command pulses will send them through whichever gate is open, when this pulse exists on the common bus line from the computer.

Since computers use one line for input-output ports, as they are called, this line might be used to feed the coder providing that any other information on the computer bus is prevented from getting into the coder by the proper gates acting in accord with the computer clock pulses. In the next chapter we will examine some ways computers are interfaced with various devices. This may be of help to us to see what and how we might interface the computer with parts of our robot.

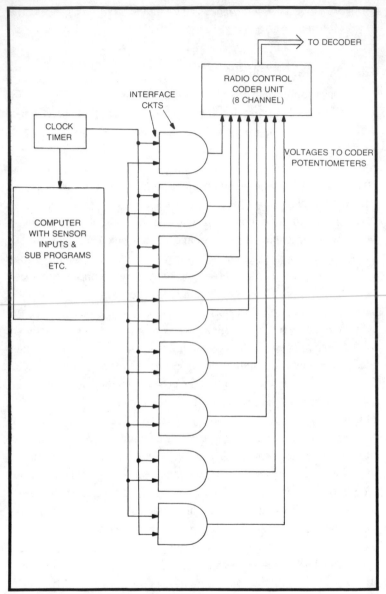

Fig. 11-6. Concept of computer addition to robot.

Having a small computer inside the robot has many advantages. First it can have a large memory. Thus we can put words there for conversations, as well as recognition information for path and position control, and action information for action of arms, motive power, security devices, etc., as necessary. The computer, as we have indicated, can be programmed either by a keyboard built internally, or by discs which simply are recorded and inserted at will, depending on what we want our robot to accomplish.

USING THE REST OF THE RADIO CONTROL SYSTEM

We might use the radio transmitter part of the radio control system inside the robot to send back information to us, or to a command computer unit located some distance away, or in the next room of the home. The transmitter input circuits can monitor information received from any source. One could have an electronic switch which rapidly samples as many as 50 points of voltage, and permits the transmitter to send out pulse positions or width variations in proportion to these voltages. At the receiving end, back where you are or the control computer is, the signal can be decoded and averaged with time so that information about everything the robot is doing and everything that is happening inside it can be known.

This is known as telemetry, the sending back of information from an object far away. It has long been used with power companies for remote monitoring of its long lines, and of course, has been used in guided missiles development, space exploration, and similar activities. The aircraft industry uses much telemetry equipment when checking out the flights of new aircraft—even the space shuttle.

So it is not a new concept either. Using it in a robot is simply adapting existing ideas and equipment to a new kind of machine. If you can call it new. We already have the transmitter if we obtained a radio control system to use, so why not utilize that transmitter portion of the system. We can use it even if we do no more than modulate it with a microphone (hidden of course) so that we can carry on a conversation with the kids down the block when our robot takes his morning stroll. We use the receiver in the robot to receive our voice signal, and the transmitter to send back the kids voices to us as we peek and monitor and reply. What fun.

There will be those who will use this bit of information as a starting point for their own developments in control and communication and to them we wish you every success. There are more possibilities in the area of robot experimentation and design than we have ever thought possible. It combines some electronics with mechanics, with chemistry, and other sciences so that one should never find it dull or feel that somehow he has reached that state where there is nothing further to try. We heartily recommend study in all these areas and thorough research on the subject as you pursue your investigations toward the ultimate cyborg.

CHAPTER 12
INTERFACING TO A COMPUTER

There are two interface situations with a computer. The input, which means adapting any number of devices to the machine to provide data or instructions, and the output from the computer which must then cause something to happen as a result of a various combination of pulses of voltage or currents. This is the end result of the computer's computing whatever it was to compute. In a sense we must speak to the machine in its language and we must convert its language output into something which we, as humans, can understand, or use.

THE REQUIREMENTS OF A COMPUTER

We have, already, to some extent, discussed some kinds of input devices which we have called sensors, and we have assumed, thus far, that the voltage or current outputs from these sensors would somehow be compatible with a computer, if our robot was such that these inputs go directly to a computing section. This is never the case. We might have an analog voltage which varies slowly and this cannot feed directly into a computer which uses pulses as its operating medium.

There are various methods by which DC signals can be converted into pulses. One method is to chop the DC signal so that it becomes a lot of voltages of different levels and then each level becomes a pulse which has some unique property such as amplitude or width or time-of-position variation according to the DC level that it was extracted from. With photocells, for example, a rotating wheel with slotted openings might be used to convert the smooth voltage level output into a series of pulses of varying amplitude, which then might be converted into the one-zero which is more usable by the computer.

A computer likes pulses which are either there (1), or not there (0), and it likes combinations of these to make up its bytes and words which give it instructions as to what to do, and how to present its output data once it has done its internal computing job. The reason the computer likes this arrangement is that it can have no confusion. If it is either there or it isn't there; there is less chance for machine error, even though sometimes when things malfunction, as they always will, a pulse might appear where it shouldn't. It is also easier to check and determine that there is or isn't a pulse in a given string position—than it is to check its width, amplitude, or time position with respect to other pulses. And if you've ever had to run down malfunctions in a computer you understand what this means.

So the first interface situation is to convert anything going into the computer into the proper pulse arrangement, and this may take some auxiliary equipment such as an analog-to-digital converter. These are manufactured as integrated circuits, or you might construct such an interface circuit using multivibrators, chopper, samplers, etc., as appropriate. It goes almost without saying that to get specific data, such as the reflections from a light beam in a robot path control system, into pulsed data for computer use, may involve quite a bit of effort, design and experimentation. Sometimes it is better to design part of the robot system as an analog system which uses, directly, the voltages obtained to control motors to obtain the required output functions.

But there are cases where the input interface is desirable. For example, it is desirable to have some computers which have a voice-to-machine interface device. James Kutsch of West Virginia University developed such a machine to enable blind persons to speak to computers directly, and thus assist them in a study of computer sciences. How wonderful a development this is— expanding the capabilities of handicapped persons who have had to rely on a sense of touch in the past. Appropriately enough the machine interface has been named Mouth (Modular Output Unit for Talking to Humans), and it also gives the computer a voice to respond to humans instructions or questions.

But this is not the total extent of experiments in this direction. Bell Telephone and Telephone Pioneers of America have developed a wheel chair which will start, stop, turn, and change speed in response to spoken commands by its individual owner. The chair does not recognize other person's voices. Here again a computer interface is required which chooses some characteristics of the human voice or breath pattern. And from this it produces voltages and currents and finally pulses which can be fed into a computer which has a comparison unit (memory). This can compare the vocal instruction to some kind of an internal instruction which can cause the energization of a small subprogram causing left turn, right turn, forward, reverse, stop, fast, and slow. Notice that there must be the comparison capability in the computer to identify that particular command with the subprogram routine desired, and cause such a subprogram routine to be initiated. Imagine also that an override must be present so that a second or third or fourth command will all be executed immediately regardless of what the chair is doing at the moment. A first order priority system relating to input must be built into the system. We have indicated that some identifying characteristic of the human voice or breath pattern can be isolated by the computer memory bank such that only that speech or breath pattern will cause the computer to activate something in its output. What are these characteristics? Well, for one thing it is the range of frequencies contained in the voice, and probably the harmonics of those frequencies. Particular patterns of the mouth and tongue may give rise to slurring of some sounds, or particular biting of words or sounds may contribute to speech individualities. The next time you talk to someone listen carefully and try to determine in your own mind what it is about their speech patterns which makes that speech identifiable with that person only.

INTERFACING MULTIPLE DEVICES

One type of interface is the analog-to-digital (A/D) converter. Another is the reverse circuit, a digital-to-analog (D/A) converter. These are important in robotics.

Although a computer is very complex and works in just a tiny fraction of a second, it is really an elaborate kind of switch which takes three signals and puts them together in such a way that only particular functions are performed in its output. These three signals are the data itself, or information, the address signal, and the control signal.

The information is a voltage level with some polarity, or a series of pulses so grouped that when they are presented to the proper devices they cause the intelligence of the output to be presented. The information may originate with sensors, from a predetermined arrangement of pulses placed on a subprogram, or from some similar source in the world of robots.

The address signal is next, and this signal insures that all needed elements are connected and those not needed are disconnected from the common bus line. Needed elements may be memories, typewriters, TVs, motors, or whatever. Then finally, the control signal (or instruction signal) makes everything work in the order and to the extent that we have imagined that it should work. The control signal causes the particular typewriter key (along with the information signal) to be activated, a particular motor to run, a light to turn on, or the printer to select a group of keys for printing.

Since the computer is a serialized machine, it feeds only one of its peripheral devices at a time. Its unit has done whatever it is supposed to do with that output device, and then it is ready to energize (in accord with instructions) the next device. The computer does this so fast, in such short time, that aside from the actual physical movement of parts in devices, we hardly realize that this is the way it is all happening. We find a similiar kind of speed in telemetry. The computer switches from one output device to another so fast that we begin to think all of them are activated simultaneously. This is important in robot operation, where many parts of the machine might have to seem to be functioning at the same time.

Of course everything that the computer does has to be referred to the computer master timing cycle. This is governed by its clock, which runs it through all its functions (whether they are performed or not) over and over again. This avoids confusion and make it possible to operate all those output devices in the proper sequence and for the required specific times without having a big signal snarl-up, which would cause overloading and jamming of the machine itself. Interface circuits are required between all parts of the computer and especially between the input sources and the output devices.

TYPICAL INTERFACE CIRCUITS

If we examine Fig. 12-1 we can recall how, in a computer, data is passed in a sequential manner from the input (buffer) devices through computer circuits to the common bus.

The buffer device may have solid-state flip-flop circuits. These flip-flops are triggered by a processor signal. At the same time the output device, which is to read and act on this signal coming from the

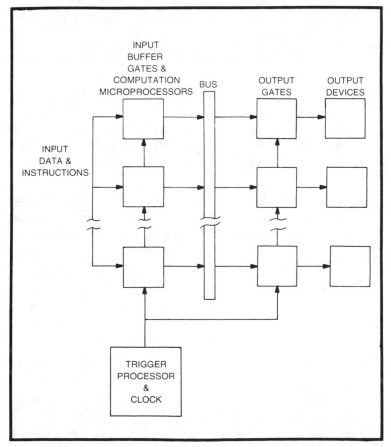

Fig. 12-1. Basic input-output configuration.

common bus must also be enabled by the trigger signal. What we seem to see here is a nice arrangement of high speed switching using a common bus line which will have signals on it coded to arrive at specific addresses of output devices which are connected to the bus for only the length of time necessary for the input signals to act upon them. This process repeats thousands of times per second.

This is all right for computer operation, where the output may be keys of a printer or an oscilloscope line (vertical and/or horizontal) or some other medium to present the data which is represented by the voltages from the input side of the diagram. The output device may actually convert these voltage levels or groups of pulses into printed characters or screen diagrams or numbers for TV display, and all this can be a very complicated unit in itself.

But in our robots we begin to think of how to make these output signals control mechanical functions, and we begin to realize that it will probably be necessary to have circuits which can control larger amounts of current and voltage than would be necessary for the printer magnets or TV timing and pulse generation circuits. We need power to run fairly large motors.

One circuit which is suggested as a means to control power in its output of the size we need is the Darlington cascaded transistor circuit which can operate a relay in its output. This is shown in Fig. 12-2.

These are important circuits for control because by proper selection of the transistors very large amounts of power can be handled, enough to be adequate for whatever action your robot may be designed to perform. You might use the Darlington (b) circuit without the relay in many cases and just let it control the power to motors. This is one of the primary buffer-to-output interface circuits used with computers for control purposes. The circuit at (c) can be a voice or DC input.

It is important to recall that there are solid state latch circuits which when set by one control pulse will remain in a closed position (giving a voltage output) until they are reset by another control pulse (making its voltage output zero). These latch circuits are made up from bi-stable flip-flop multivibrators.

If a relay is placed in the proper output of such a latch circuit it will remain energized until it is reset. It may be necessary to use the

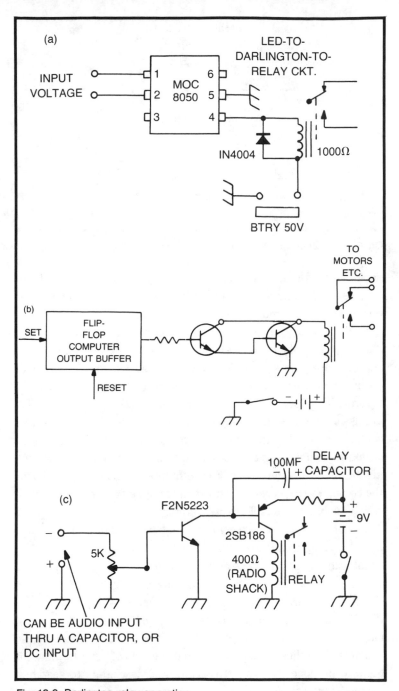

Fig. 12-2. Darlington relay operation.

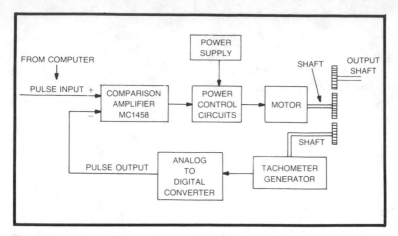

Fig. 12-3. An analog-to-digital circuit for controlling the speed of a motor.

sensitive reed-type relays for use in this kind of application and then let these relays operate larger power relays for the application of the power needed for motors or other such devices.

Because a robot application will have motors of various kinds, it may be necessary to have some means of using the computer to control the motor speed. We have already shown in a previous chapter how we might use a tachometer feedback to give us more accurate positioning of a motor driven output arm, and in that discussion we used the voltage generated as an analog input to a control amplifier. Now we consider the possibility of using an analog-to-digital integrated circuit which then might compare the number of pulses generated, to the number of pulses called for by the computer to get very accurate motor speed control. One circuit which might do this is illustrated in Fig. 12-4.

Of course the reverse might also be required. If you have pulses coming from the computing section of the robot brain you might want to convert these into an analog voltage which will specify an input command to a servomechanism as described in an earlier chapter. That arrangement is shown in Fig. 12-4.

You can always find circuits which will give you the kind of interface you need to work with the kind of input or output signals you must feed them, and they produce the kind of outputs you want them to produce. One way to locate such circuits is to search in current literature. Another is to write to appropriate manufacturers and ask them about the circuit they may suggest to use with the devices on each end of the interface that you plan to use. We are not hedging the question, but know from experience that time will produce many changes in this area of circuitry, and so believe it best to let you check with current concepts of interfaces as you need them.

DESIGN APPROACHES TO A COMPUTERIZED ROBOT

It is always a good approach in designing a computerized robot to get out your pencil and paper and put down all the functions which you ultimately want performed. You may not know how you are going to accomplish each one at this time, but you do need to have each and every function listed that you can possibly think of. They do not have to be listed in any particular order. They just have to be listed. That is the first step. Let your imagination run wild and put down everything. You may want to eliminate some things later and you may combine functions, but you do need to have everything you can think of down on this first big list. In the commercial world this is known as brainstorming, at home get the family's ideas.

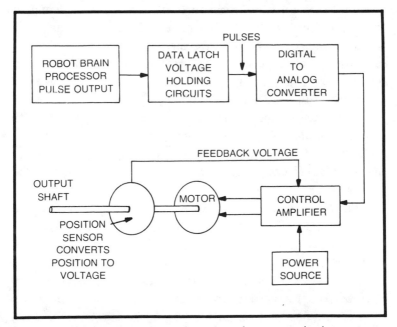

Fig. 12-4. Converting pulses to analog voltage for servomechanism.

Next you (and your associates) begin to analyze each function separately to see just what will be required in the way of motions or other phenomena. You will also look at the dependence or interdependence of these things, one with respect to all others. Then as you analyze the conversion of light, heat, sound, and radiation, to signals which will be used in the machine then you will begin to think about how the machine is going to work to produce the required outputs.

When you do this thinking about the how, you will begin to come up with the kind of brain the robot is going to need. And it may be a computer in the sense that we now think of computers, a microprocessor, or just transistors and integrated circuits. Whatever computing is done will be done only when and where it is appropriate by whatever addition and subtraction or latch or whatever kinds of circuits are needed to control the main power flows in the machine.

If you come up with the centralized computing section then you will have specified the requirements for a brain as we now think about it. If it is all in one place (like ours) and it is designed so that it can be modified, expanded, repaired, or adjusted, then it can make the construction and maintenance process easier. We do not rule out the possibility that the robot might be given the basic means of repairing and adjusting itself. Was that on your list?

With a brain it might be easier to give your robot new learning capabilities because you can add more or better or different memories as they are developed and as you find the need for them.

You teach a robot by first having it perform the actions you will want it to perform alone, by sending it such signals as it will require to do these jobs. It then memorizes these commands. It also memorizes the feedback signals which indicate that it has done what it was commanded to do.

In many instances this means of putting the input signals into a robot in industrial applications is done by maneuvering the robots with a human attachment, an arm for example. This can grasp and move, which causes the robot arm to move in exact compliance with the way you move the teaching arm. Signals are thus generated, memorized, and stored in the robot's memory. After you have taken it through its paces it can then repeat these moves in the exact manner and in the exact sequence in which you made the movements first. Teaching (or some call it programming) a robot is and will continue to be one of the more important capabilities you must have with regard to your machine if it is to do any useful work.

This teaching step might be called the beginning step of robot operation. It will require a human-to-robot interface of mechanical and electrical attachments which will be removed once the teaching process is completed. This human-to-robot interface might have to be more than a voice processor, depending upon what you have listed as the robots tasks and duties and jobs. Some say that it will only be necessary to feed the information from the interface to the robot computer. We doubt if this will always be sufficient.

Interfacing devices are not only those which connect the robot to the outside world, they are also those devices which connect the various internal parts of the robot to other internal parts in order to cause a smooth and accurate functioning of the device as a whole. That makes three basic areas which are important in robot design. The robot thinking, or analyzing, processes and equipment, the robot mechanical equipment, and the interfacing equipment. As you plan your own robot you will think much and long about each of these areas.

CHAPTER 13
MODERN ROBOTS
AND ROBOT SOCIETIES

More and more countries throughout the world—Japan, the Soviet Union, Italy, France, Britain and the U.S. to mention the leading ones—are working feverishly trying to improve upon industrial robots.

These are not only the robots we examined earlier, those that can explore the ocean floor, or work in nuclear "hot" areas, or handle masses of materials too unwieldy and nasty to be handled by humans. No. These countries are trying hard to develop robots for the thousands of assembly lines filling our factories. Japan has under test as of this writing a factory which is completely automated by robots. These are fixed in position and work according to a master plan to assemble equipment for the markets of the world.

The goal set by these countries is to gradually replace the human robot who sits day after day doing the same task in exactly the same manner hour after hour. Granted that there are thousands of people who like this atmosphere and work. They do what they have to do very well, and they have no desire to always have to face the task of finding solutions to new problems or new situations which arise in other kinds of work. These persons like the security of knowing their job, doing it well, and receiving good pay for their efforts.

PROBLEMS WITH AUTOMATED FACTORIES

If we consider what must be accomplished by the machine sitting in place of a human on an assembly line, then we begin to get an appreciation of the problems involved in automation. For example, if parts are placed in the front of the bench position and ten of these are to be assembled at that bench position, and the parts lie in a random array inside of bins, we have a job which requires the human ability to select, turn, feel, and see. And then after grasping the part, to place it into the proper holes or on the proper terminals or in some other precise position and securing it there in the proper manner.

If you consider a robot arm which has a hand able to grasp small parts, then you see that the first problem is that the hand can select from all the parts in the bin only those which might have a particular orientation with respect to the hand-fingers arrangement of the robot. Of course the hand might be caused to stir the mix and then make a selection, but this also might result in broken parts or a worse arrangement of the parts than they had in the first place.

How about giving the robot eyes so that it can see the arrangement, just as a human does? This, of course, is possible. The problem here is that you would have to provide the computer memory, which obtains the information from the eyes, with every possible arrangement of parts, and then provide the computer with some means of determining which one of those the hand should try to pick up first, second and so on. Also it must decide how to position the hand to get those particular parts when they are needed.

If you are thinking along with us, then at this moment you are probably saying that we should not use bins of parts, but instead place all parts on some automatic conveyer feeding belt so that only one at a time is presented to the robot arm and each has the right orientation. Then the hand can grasp it and move it to the proper location on the assembly chassis. True! And this might be a workable solution. Now let us consider that a part has been selected and the robot is to fit it to some particular terminal or hole or fastener.

It was shown by personnel at the Draper Laboratories, Cambridge, Mass., that the actual fitting of a part into a hole involves more than just trying to get it into the proper spatial orientation. It seems that we humans, when fitting parts in this manner, may jiggle them into their holes or slots, and with this jiggling or wiggling of the part, do make it fit in spite of friction or a slight binding or whatever may occur when the fit is not precise. Now the problem here is amplified if the tolerances are small and the fit must be tight.

In an arrangement which may involve such mechanical assemblies as fitting nuts on bolts, fitting parts into slots or keyways, and installing parts on printed circuit boards, putting parts exactly where they belong when the chassis becomes crowded produces great difficulty. To fit wire terminals around a terminal strip as performed in some electronic assemblies requires a special maneuvering capability for a robot hand which is different from the movement needed in the general assembly of parts.

In a study conducted by the Cambridge group under a National Science Foundation grant and illustrated in the February, 1978, issue of "Scientific American," it was found possible to completely assemble an alternator for an automobile using a single, instrumented robot arm. This arm was capable of changing its hand when different tools were needed in the assembly process. Different hands had different tools built into them. Another concept that made this fabrication possible was that all parts came from one direction and maintained a specific orientation on the feeder belt.

It has been shown in some other studies that the requirement for use of coordinated hand-arm movements is a much more difficult process in the mechanization development. This movement occurs when assembly may have to be from two or more directions and some movements are done, say, on the bottom of the assembly, while others are built up on the top side of the assembly.

The Draper group found that giving a robot eyes did not improve its capability to coordinate the use of both arms and hands. They found that the movement of the robot arm alone in a precise and exact manner to reach exact spatial coordinates was far more important than trying to let the robot see what it was doing. Coordinated movement of both arms and hands would become a marvelous computerized arrangement and be directed by a computer that would be fantastic.

We have always held the fondest memories for the work of Charles Stark Draper and hold him in highest respect. He invented the inertial guidance system for guided missiles and for other applications. That system relies on spatial coordinates and accelerations in these coordinates to give spatial position very precisely.

So we begin to understand something of factory automation problems. Somewhere along the line, certain knowledgeable wizards will tell us that this isn't the only problem to reckon with. They will tell us there is a cost-capability factor which must be taken into account. It might be possible now to automate an entire factor, but the cost may be so high that it just wouldn't be an economical effort.

MAINTENANCE AND REPAIR VIEWPOINTS

Once you have constructed a small robot or automated machine of any kind and begin to operate it you begin to find out about the maintenance (preventive and actual) required to keep it operating as it should operate. You then become aware of another problem in factory automation. If you have a thousand bench positions, each having a robot arm which is highly instrumented to perform its part of some fabrication, you realize that any small deviation at any position of any part of the system can cause delays or foul ups. These cost money.

The systems must be so designed that each vital part of the individual robots must have some fault indicator and safety mechanism, the first to show that trouble is developing, the second to shut things down so that repair and adjustment can take place before anything serious happens to the whole factory. This maintenance and repair requirement will probably generate a whole new science and will require a whole new breed of engineers and technicians. These men must be experts who can perform their difficult and demanding tasks quickly and accurately.

Yes, there is a big effort now going on all over the world to develop robots which can do factory assembly jobs fast and sure and do them without undue machine complexity and with the highest degree of reliability possible. It may turn out that dual computer systems will be needed. After all, it has been said that we humans have dual capabilities in our brains, two halves of which can perform the same functions.

If we have this redundancy, then it seems reasonable to assume that any mechanical-electronic robot should also require a certain amount of duplication of parts and control systems. There should be indicators to show when one system part is off tolerance or is failing. Then, at the same time, it will shift to the redundant set of equipment or instrumentation to keep that total machine going while the first part is being required or readjusted. Self repair, by the robot in such cases is a current dream.

It is informative to know that the financing of the investigation of manufacturing techniques by robot machines is being done by the National Science Foundation and also some industrial firms. The studies show that now it is feasible to use some automated machines on some assembly lines even today. It has been found that wrist-hand movements having compliance like a human wrist-hand action has made it possible to predict that now even small quantities of items might be assembled by such robots. This can be done with speeds and with such precision and at such costs that these projects could have a gainful financial reward.

Thus while we may not see manufactured home robots of the walking and talking type in the near future, depending, of course, on the success of such firms as Quasar industries attempts to make such machines, we may indeed find a vast increase in the varieties of robots used in industry. The improvements in the computer and microprocessors will give added impetus to this development, and it is to them that we begin to look for technological breakthroughs. The actual techniques of electrical-to-mechanical movements using servo-mechanisms are at this date pretty well known and under-stood, even to the types called adaptive which tend to learn by their own mistakes. But there could be technological breakthroughs in this area also, and of course, in sensors.

ROBOT SOCIETIES

Yes there are some, both industrial and hobby types. In America there is an institute called the Robot Institute of America, which is concerned with the industrial types of robots. We made some contacts with this society in our exploratory effort to find out as much about robot use, development, and planning as possible. They gave us some information on what they do and how to join such a society to gain such information as they get and pass onto members. They are internationally recognized and so cover the world in this respect. They welcome members and so we include a sample of their membership application form for your information in Fig. 13-1.

We enjoyed nice conversations with Mr. Donald A. Vincent, manager of the institute and learned that there are over 400 current members. These members come from some of our best known and respected agencies.

There are many hobby-robot societies coming into being as the home computer gains recognition and more wide-spread use. If you are interested in this area—which a lot of us are—you might contact the International Institute for Robotics, Box 615, Pelahatchie, Mississippi, 39145. There is no question that as time goes on more and more builders of hobby robots will want to challenge other robot builders to various contests involving capabilities and tech-niques. What fun that could be.

We have found in our investigations that in a scientific sense, many agencies prefer to use the word manipulator rather than robot when referring to the automated machine. They believe that the word robot has a connotation toward a semihuman creature. We have found in the literature that the word automation and man-ipulator is frequently used while there seems to be a tendency to avoid the use of the word robot. Be guided by this as you continue with your own research.

SOME CAPABILITIES AND TYPES OF INDUSTRIAL MANIPULATORS

Let us examine for a few moments some other machines being used, developed, or under investigation by such personnel as make up the Robot Institute of America. One type of (we're going to call them robots) robot is an automatic paint spraying machine which is produced by the Binks Manufacturing Co. It is said to have six degrees of freedom and can be movable or fixed. It has feedback sensors for adaptive control and thus can duplicate the actions of an expert human paint sprayer, once it has been taught the procedure. Perhaps you wouldn't think of an automatic paint gun device as being a manipulator or robot, but it does fall into the definition which fits this kind of machine.

Many of the manipulators similiar to this include features like loading and unloading of materials. These are programmed by step-ping switches, tapes, or discs which, as we earlier indicated, can

MEMBERSHIP APPLICATION

® Robot Institute Of America

20501 FORD ROAD, P. O. BOX 930, DEARBORN, MICHIGAN 48128, (313) 271-1500

—PERSONAL AND BUSINESS INFORMATION:

Check Preferred Mailing Address

☐ HOME

☐ BUSINESS

Name _____ Last Name _____ First Name _____ Middle Name _____ Date of Birth _____

Home Address _____ No. & Street _____ P. O. Box No. _____

City _____ State _____ Zip _____ Home Phone _____

Company Name _____

Division/Department _____

No. & Street _____ P. O. Box No. _____

City _____ State _____ Zip _____ Company Phone _____

Job Title _____ Major End-Product or Service _____

Job Function:
☐ Company Management
☐ Manufacturing Production
☐ Engineering
☐ Product Design/ R &D
☐ Marketing Sales
☐ Other Explain _____

Plant Size: What is the Size of Your Plant by Number of Employees? Please Check Correct Number
☐ Less than 50 ☐ 50-99 ☐ 100-249 ☐ 250-499 ☐ 500-999 ☐ 1000-2499 ☐ 2500 and Over

—ADVANCED EDUCATION:

Technical Institute or College Graduate: ☐ Yes ☐ No

Years Completed (Circle): 1 2 3 4 5 6 7

Degree Granted: ☐ None ☐ Associate ☐ Bachelor ☐ Master ☐ Doctorate

Field: ☐ Engineering Technology ☐ Engineering ☐ Science ☐ ______ Other Specify

—PROFESSIONAL STATUS:

☐ Registered Professional Engineer ☐ Certified Manufacturing Engineer ☐ Certified Engineering Technician ☐ Engineering in Training

—GRADES OF MEMBERSHIP:

☐ Individual Members $43.00
($28.00 yearly dues plus
onetime $15.00 initiation fee)

☐ check enclosed ☐ bill company ☐ bill me

Applicant's Signature ______ Date ______

FOR HEADQUARTERS USE ONLY:

Date ______

Dues Amount ______ PROCE. ______

Init. Fee ______ Member No. ______ DUNS NO. ______

Total ______ DATA PROCE. ______

SIC	Company	Title	RE	Education	Verification Mo.-Yr.	Code Source	Plant Size	Company Address City	Company Address State	Residence—If Preferred City	Residence—If Preferred State

Fig. 13-1. Robot Institute of America Membership Application.

furnish programs for home robots. The memory drum we examined in a earlier chapter is an elementary unit which, when made more sophisticated by using switches and perhaps cams of special shapes, does the job which we sometimes think can be done only by microcomputers.

There are robots from Modular Machine Co. of San Diego, California, which are cam controlled. They use cams on a program drum to operate snap-action switches. These are called vectrons. It is said that several of these can be custom assembled for linear and rotary motions and plugged together end to end to give compound motions. Repeatability is said to be 0.005 inch and position accuracy plus or minus 0.005 inch. These vectrons have a limit switch at each end of travel which is interlocked with its controller. New motion commands are issued after a limit switch signals back that the preceeding command has been fully obeyed. The motion commands are issued by a stepping switch in the controller which responds to the limit switch signals. This is a full movement type machine.

The Overton Engine Co. of Harrisburg, Pa., offers a modular approach to the industrial manipulator. These modules are the transfer mechanisms which cause arms to move in a fixed path cycloid motion with a shockless, smooth acceleration from zero to final velocity. The paths are gear generated. With the modules available you can construct your own robot-manipulator to do whatever job you have in mind for it to do.

Pick-O-Matic Systems, Inc., which is a division of Fraser Automation of Sterling Heights, Michigan, has a mechanical parts handler which may perform either a single motion or a double pick-and-place motion. Optional units have four axis. These units are said to be compatible for adaptive control type operations.

Robmation Corp, Southfield, Michigan, provides a complete parts handling robot with sequence programming, or robot components which you might desire for your own robot system. Standard slide and rotary actuators combine into a small-parts handling robot-manipulator. Building blocks said to be available are the Robo-Slide for linear motion, the Robo-Rotor 180 degree rotary actuator and the Robot-Grammer sequence programmer. These systems are hydraulics or low pressure air and you can assemble them to get up to five axis of motion with speeds of up to 600 inches per minute. The cams in the programmer are said to be easily replaceable for new and different operations.

Industrial Automates, Inc, Milwaukee, Wisconsin, offers a solid-state robot for under $10,000. This unit is said to have many of the features of the big, more expensive industrial robots, and retains all the advantages of those systems. The basic operation is similiar to a conventional stepping switch except that each line of its 128 step RAM memory is dedicated to a single piece of information. Using the high speed stepping characteristics of the system, a series of separate commands or outputs can appear as though they occurred simultaneously—or you can have time intervals inserted between the commands to suit your demand requirements. The communication method bypasses the extensive ladder diagrams of Boolean algebra or similar languages. The simple code written into the memory of the 16/12 unit is produced while stating the sequence of events for the robot program.

Then as each line in the memory is addressed, two rotary switches select a number and a letter, the write button is pressed, and you step to the next line. You can write an average program in this manner in about four minutes and the program consists of about 28 steps. If you make a mistake (heaven forbid!) you can remedy the situation by readdressing the memory to the correct line setting the correct code and pressing the write button. The new data automatically erases the old data.

The Cincinnati Milacron Co. of Cincinnati, Ohio has developed an arm called the 6CH arm which is a general type manipulator capable of performing delicate, precise tasks such as wire harnessing or arc welding and can handle loads from 175 to 300 lbs under limited conditions. It is designed for rugged duty operation or repetitive tasks in dangerous environments and for exacting tasks than exceed human capability under these conditions. The arm is mounted on a large base and has all the required movements of horizontal sweep, vertical shoulder and elbow swivel, with wrist pitch, yaw, and roll. The arm has six axis motions anywhere throughout a 1000 cubic foot envelope approximating a 16 foot hemisphere. The arm uses hydraulic motors and has a computer controller.

The question for vision in a machine, and its usefulness, is said to be demonstrated by Auto-Place, Inc., of Troy, Michigan. They offer robots with "vision." They believe that the areas of inspection and quality control are suitable for this kind of machine.

Many ordinary operations which require visual observation but which are tiring to human personnel might be performed satisfactorily by this kind of robot. Recently, there have been some rather simple techniques employed to give robots a sense of vision. We have described one method earlier in this work. The Auto-Place limited sequence robots are said to use both laser and video camera technology to provide the vision needed to handle small parts. It is said that this Opto-Sense, as it is called by Auto-Place Mfg. Co., has made it possible to use the machines in gauging, inspection, sorting and similiar type tasks which a robot without vision might not be able to accomplish. Here is how they are said to operate.

When using laser beams, which are small and very precise, and needlepoint, the robot can scan the part or work, and then compare that part or work to a master model of the product. If parts are missing, or holes not drilled, or not drilled correctly, then the part being inspected can be rejected. The use of video scanning techniques will send the data to a computer which makes the electronic comparison necessary to accept or to reject the part or assembly. So the use of a seeing robot in industry is a fact and its use will undoubtedly be expanded.

George Munson of Unimation, Inc., Danbury, Conn., specializes in complete systems such as those used to load bricks into kiln cars. These same type systems of automation apply to auto manufacturing plants where many processes, such as assembly, transportation of parts, and selection of parts, have to be accomplished.

From Programmable Automation, Westinghouse Electric Corporation, Mr. Abrahams, the manager has said that lower cost industrial robots will be the key to increased production, especially for batch assembly. Batch assembly is characterized by variations in product styles that require frequent changeovers and set-ups. The parts may vary in geometry also from batch to batch. He has also said that very precise motions and movements are required, and that the robot might have to adjust its path of operations to accommodate for variability in parts, as things have small changes. The robot must have some kind of adaptive control to permit it to adjust to small changes which ultimately, with time, may mean large changes from its initial operation.

MORE ABOUT THE ROBOT INSTITUTE OF AMERICA

Not only does the Robot Institute of America have much general information on industrial robots, they also have printed papers which have been given at the many symposiums held all over the world on the subject of robotics. They can provide you with information on where and what papers will be given in the future on this subject. Some of the titles which reveal the scopes of the papers are as follows:

- "A Users Guide to Robot Applications."
- "Designing Robots for Industrial Environments."
- "Practical Applications of a Limited Sequence Robot."
- "Application Flexibility of a Computer Controlled Industrial Robot."
- "A Multidimensional System Analysis of the Assembly Process as performed by a Manipulator."
- "Cost Effective Programmable Assembly System."
- "Artificial Sensing."
- "Interface and Control Strategies for Industrial Automation Systems."
- "Computer Control of Robots—A Servo Survey."
- "Software Features for Intelligent Industrial Robots."

These papers may be obtained from the Society of Manufacturing Engineers or the Robot Institute of America for various costs. There are also many films about Robots which are available, for rental fees, from the institute, and from various manufacturers which the institute can put you in communication with. It could be worthwhile, if you are really a dyed-in-the-wool robot enthusiast, to contact the institute and get all the information in that particular phase of robotics which might be of the most benefit to your future ambitions in this field. It is a growing and fast developing field, and as we have stated, there are now agencies all over the world deeply involved and committed to the development of these machines.

Almost every possible subject matter associated with robotics has been presented as a paper and what we look for as time goes by is the application of new techniques and new technology and breakthroughs in technology which will make the past-impossible become possible. Much is to be said for mathematical study, and much must be said for the experimentation of the home experimenter, for this person doesn't know what is impossible and what is not, and often times does the impossible just because he didn't know it wasn't supposed to be possible to do.

THE INDUSTRIAL ROBOT JOURNAL

We would be remiss in providing you with all possible information about robots and where and how to get more information on this subject if we did not mention the trade journal of robotics, called the "Industrial Robot" and printed in England. You can obtain information on this publication which purports to keep up with all current developments in this field, by contacting International Fluidics Services, Inc., 35-39 High Street, Kempton, Bedford, England. This publication is quarterly and follows world wide developments in the robotics field. It could be worthwhile to obtain copies as the robot field is fast moving, and the technological advances are incredible. What seemed impossible yesterday is easily accomplished tomorrow.

CHAPTER 14
THE FUTURE OF ROBOTICS

Although the robot has always been a fascinating subject and as of this writing produces vision of those used in such movies as "Star Wars, "Close Encounters," and "The Day The Earth Stood Still" and even more fantastic kinds are as drawn in comic strips and alluded to in science fiction books, we have found that our modern robots even seem to outstrip these in reality, capability, and development. What has and is being accomplished is truly incredible.

PAST AND FUTURE

Looking into the future makes us think also of the past. Some will remember when radar first controlled anti-aircraft guns, when the first pilotless aircraft came into being, and when the subject of servo-mechanisms first appeared on the list of college technical subjects. The subject of computers, which also began its development at that time—first in the form of complicated mechanical devices which "solved" the ground-to-air gunnery problem and progressed to the vacuum tube giants and relay computers which made so much noise that you wondered if the problems solved were worth that much sacrifice of peace and quiet—began to grow and become a separate subject itself.

From our examination and diligent prying into this field, we now believe that anything is possible. Humanized robots are possible—if a demand for them exists. What they can do around the home, office, or other place of business remains to be envisioned in some active and imaginative brains of humans who have not learned to think in terms of "not practical" or "not possible." These minds, adventuresome and eager, will take the thousands of off-the-shelf items, some of which we have described and others which we have alluded to in this work, and will fabricate them together with new and wonderful additions to create more marvelous inventions and adaptations than we have ever dreamed of. So let it be that the future of robotics belongs to the young and eager and the young-at-heart.

We now believe that the only reason we do not have *R2-D2's* and *Threepios* and still more advanced cyborgs is not because they cannot be developed, but because there has not been a need specified for them. As that need becomes evident, then they will appear in multitudes and we predict they will ultimately perform most all household tasks. Won't the ladies love that.

It seems there is still a large question as to the form which the home robot should take. Should it be simply a built-in computer controller. One that through many tentacled electrical arms and control devices causes things to happen around the home, things such as garage door openings, lawn mowing, hedge trimming, cooking, dish cleaning, washing, inventories, selective heating and cooling (which means you don't heat or cool the rooms you don't need to) and TV program selecting (so the kids won't be able to get programs not good for them). Or should the modern robot be our always present TV sets which present the games and learning and talking ability using some centralized computer of giant size which may be a part of our telephone complex and which brings us the finest in library contained sciences, arts, music, movies, instruction, discussions—in fact anything we have a desire to learn or see or hear or know about. We simply select by voice or button what we want and it is there. And if we don't know what we want—there may be a provision for that also. The computer will determine our mood and desires and find something appropriate for us.

Or will it be the actual robot companion such as we have seen and described in earlier pages. It could be, and it could be any day now that we desire such a companion. Of course the capabilities of units the (and with Kryton and Klatu staring at me) must be increased. But then, doesn't it take some 15 to 25 (or more) years to make an intelligent, thinking, responsible human being out of a baby? You might also add the phrase "acting human beings" to the previous concept. The robot must act as well as think intelligently. What is needed depends upon what we really want the humanized robot to do. And you must think that out. Someone must come up with all the possible tasks which make sense, and then, from that list will come the development and acceptance of robots, not as toys or a creative and inspiring marvel to be seen and wondered about and forgotten, but a machine which everyone will agree belongs in the well organized and regulated and sensible household.

This is not to say we should not experiment. That we must do. And we must do it to the utmost of our ability. Once we have experimented with fundamentals and basic machines, then, perhaps we will study and advance to the next level of scientific knowledge and experiment some more, and with more sophistication, and toward more purposeful goals, and we will study some more and experiment some more *ad infinitum*. Ultimately we progress from the toy or fun stage to something which does have an answer to the question so many ask "What good is it?" Then we can reply "What good is a baby?"

There is so much development and advancement in the field of industrial automation and robotics that we can only say that a greater exchange of ideas and more discussion of what is being done at agencies all over the world should be made available to the youthful and advancing future scientists of the world. It is hard to get current information as mercenary considerations arise about patents and manufacture and operations and breakthroughs. It is also hard to get information on such subjects (because of complexities of technical reporting) until one approaches college age and does a thesis or other paper on this area of development. We suggest that youthful perons investigate the extensive literature in college and university libraries as part of their effort to keep abreast of this exciting field of automated engineering and technology. Local libraries might be able to provide indexes of subject matter if they do not have the material itself available.

So the world of robots is real and fascinating and in its infancy. We have tried to cover the subject from toys to historical types, to current types of robots to include an assessment of the considerations of robots as to what is available and possible (at this time) and what is not. We have made an effort to show and discuss various types of robots now in existence and to indicate briefly how they perform. We have also tried to point the way toward further studies which you might want to attempt. What do we say in closing? We thank you again for being with us on this journey. We really do hope that you will find the information useful and of value to you, and we hope above all, that your imagination will be stimulated such that you will continue to investigate and experiment with this new area of technology.

All the good numbers to you and yours, and as Kryton always commands "Keep your gears oiled."

INDEX

BELMONT COLLEGE LIBRARY